# Bridging Quantum and Classical: A Path Integral Journey

Chopra

# TABLE OF CONTENTS

# Chapter 1

# CAYLEY MODIFICATION FOR PATH-INTEGRAL SIMULATIONS

## 1.1 Introduction

Feynman's path-integral formulation of quantum statistical mechanics[2] offers pow-erful and widely used strategies for including nuclear quantum effects in complex chemical systems. These strategies are based on the observation that the quantum Boltzmann statistical mechanics of a quantum system is exactly reproduced by the classical Boltzmann statistical mechanics of an isomorphic ring-polymer system.[3] For the numerically exact calculation of quantum Boltzmann statistical properties, the classical Boltzmann distribution of the ring-polymer system can be sampled using Monte Carlo[4] (i.e., path-integral Monte Carlo, or PIMC) or molecular dynamics[5] (PIMD).

For the approximate calculation of dynamical quantities, such as reaction rates,[6–8] diffusion coefficients,[9,10] and absorption spectra,[11–14] the Newtonian dynamics of the classical isomorphic system can be numerically integrated as a model for the real-time quantum dynamics, as in ring-polymer molecular dynamics (RPMD)[15,16] and centroid molecular dynamics (CMD).[17,18] These and related methods have enjoyed broad applicability in recent years for exploring nuclear quantum effects in the domains that span physical, bio-, geo-, and materials chemistry.[1]

For PIMD and RPMD calculations, considerable effort has been dedicated to the development and refinement of numerical integration schemes. This work falls into two distinct categories. In the first, the RPMD equations of motion are preconditioned by modifying the ring polymer mass matrix; this causes the integrated trajectories to differ from those of the RPMD model,[19–25] but it can lead to efficient and strongly stable[21–23] sampling of the quantum Boltzmann distribution. In the second category, no modification is made to the ring-polymer mass matrix (i.e., the "physical" masses of the ring-polymer beads are employed).[26–30]

Within the second category, it is common to apply a thermostat to the internal ring-polymer motions, with two primary aims: to more efficiently sample the quantum Boltzmann distribution,[24,27,28] or to avoid the "spurious resonance" artifact of the microcanonical (i.e., un-thermostatted) RPMD equations of motion in which internal ring-polymer modes mechanically couple to physical modes of the system.[29,30] PIMD and RPMD integration schemes in the second category (which preserve the RPMD model dynamics) typically employ a Trotter-like factorization of the time evolution operator.[9,27–31] For the example of thermostatted RPMD (T-RPMD)[29] using the generalized Langevin equation (GLE) thermostat,[27] the numerical integration is performed using[32]

$$e^{\Delta t L} = e^{\frac{\Delta t}{2} L_\gamma} e^{\frac{\Delta t}{2} L_V} e^{\Delta t L_0} e^{\frac{\Delta t}{2} L_V} e^{\frac{\Delta t}{2} L_\gamma} + O(\Delta t^3) \tag{1.1}$$

where the Liouvillian $L = L_V + L_0 + L_\gamma$ includes contributions from the physical potential, $L_V$, the purely harmonic free ring-polymer motion, $L_0$, and the friction and thermal noise, $L_\gamma$; note that the standard microcanonical RPMD numerical integration scheme is then recovered in the limit of zero coupling to the thermostat, such that[9]

$$e^{\Delta t L} = e^{\frac{\Delta t}{2} L_V} e^{\Delta t L_0} e^{\frac{\Delta t}{2} L_V} + O(\Delta t^3). \tag{1.2}$$

Standard practice in these RPMD and PIMD integration schemes is to exactly evolve the harmonic free ring-polymer dynamics associated with $\exp(\Delta t L_0)$ using the uncoupled free ring-polymer normal modes.[9,27,31]

The first major conclusion of this Chapter is that any integration scheme (PIMD, RPMD, CMD, or other)[5,15–17,27–31,33–38] that involves the exact integration of the free ring polymer (i.e., involves the ubiquitous $\exp(\Delta t L_0)$ step in terms of the ring-polymer normal modes) will exhibit provable numerical deficiencies, including resonance instabilities and non-ergodicity. For the case of the standard microcanonical RPMD integration scheme in Eq. (1.2), which is a symplectic map, exact evolution of the free ring-polymer step leads to the provable loss of strong symplectic stability and the demonstrable appearance of resonance instabilities in the integrated trajectories. For thermostatted RPMD and PIMD integration schemes that involve a free ring-polymer step,[27–30] exact evolution of that step leads to the provable and numerically demonstrable non-ergodicity.

The second major conclusion of this Chapter is that these numerical artifacts can be eliminated by simply replacing the exact evolution of the free ring polymer step with an approximation based on the Cayley transform: an alternative to exact free ring-polymer evolution that is no more costly, no more complicated, and no less accurate in the context of the full integration timestep. In particular, we show that this Cayley modification eliminates the resonance instabilities that occurs when trajectories are evolved using standard microcanonical RPMD integrators, and we show that it restores ergodicity to thermostatted RPMD and PIMD trajectories. Furthermore, we show that the improved numerical properties of the Cayley modification generally allow for larger RPMD and PIMD integration timesteps to be employed.

The Chapter is organized as follows. In section 1.3 we articulate the numerical instability problem in the context of standard RPMD numerical integration and introduce the Cayley modification as the solution. Section 1.4 numerically illustrates the instability of standard RPMD numerical integration and shows that the Cayley modification removes this problem. Finally, in section 1.5 we generalize these findings to thermostatted trajectories.

## 1.2 Theory

The theory introduced here adapts and advances previous mathematical results on the numerical approximation of general second order Langevin stochastic partial differential equations with space-time white noise.[39]

## RPMD

We consider a quantum particle in 1D with Hamiltonian operator given by

$$\hat{H} = \frac{\hat{p}^2}{2m} + V(\hat{q}) \tag{1.3}$$

where $\hat{q}$, $\hat{p}$, and $m$ represent the particle position, momentum, and mass, respectively, and $V(\hat{q})$ is a potential energy surface. All results presented here are easily generalized to multiple dimensional quantum systems.

The thermal equilibrium properties of the system are described by the quantum mechanical Boltzmann partition function,

$$Q = \text{Tr}[e^{-\beta \hat{H}}] \,, \tag{1.4}$$

where $\beta = (k_B T)^{-1}$ is the inverse temperature. Using a path-integral discretization, $Q$ can be approximated by a classical partition function $Q_n$ of a ring-polymer with $n$ beads,[5]

$$Q_n = \frac{m^n}{(2\pi\hbar)^n} \int d^n q \int d^n v \, e^{-\beta H_n(q,v)} \,, \tag{1.5}$$

where $q = (q_0, \ldots, q_{n-1})$ is the vector of bead positions, and $v$ is the corresponding vector of velocities. The ring-polymer Hamiltonian is given by

$$H_n(q,v) = H_n^0(q,v) + V_n^{\text{ext}}(q), \tag{1.6}$$

which includes contributions from the physical potential

$$V_n^{\text{ext}}(q) = \frac{1}{n} \sum_{j=0}^{n-1} V(q_j) \tag{1.7}$$

and the free ring-polymer Hamiltonian

$$H_n^0(q,v) = \frac{m_n}{2} \sum_{j=0}^{n-1} \left[ v_j^2 + \omega_n^2 (q_{j+1} - q_j)^2 \right], \tag{1.8}$$

where $m_n = m/n$, $\omega_n = n/(\hbar\beta)$ and $q_n = q_0$. If we let $n = 1$ in Eq. (1.5), the classical partition function of the system (governed by a classical Hamiltonian, Eq. 1.6 with $n = 1$) is recovered, i.e., $Q_1 = Q_{cl}$. In the limit $n \rightarrow \infty$, the path-integral approximation converges to the exact quantum Boltzmann statistics for the system, such that $Q_\infty = Q$. The thermal ensemble of ring-polymer configurations associated with Eq. (1.5) can be sampled using either molecular dynamics (leading to PIMD methods) or Monte Carlo (leading to PIMC methods).

The classical equations of motion associated with the ring-polymer Hamiltonian in Eq. (1.6),

$$\dot{q}_j = v_j, \tag{1.9}$$

$$\dot{v}_j = \omega_n^2 (q_{j+1} + q_{j-1} - 2q_j) - \frac{1}{m} V'(q_j),$$

yield the RPMD model for the real-time dynamics of the system.[15,16] RPMD provides a means of approximately calculating Kubo-transformed thermal time-correlation functions, such as the position autocorrelation function

$$\tilde{C}_{qq}(t) = \frac{1}{Q} \text{Tr}[e^{-\beta\hat{H}} \tilde{q}(0)\hat{q}(t)] \tag{1.10}$$

where the Kubo-transformed position operator $\tilde{q}$ is

$$\tilde{q} = \frac{1}{\beta} \int_0^\beta e^{\lambda \hat{H}} \hat{q} e^{-\lambda \hat{H}} \, d\lambda \qquad (1.11)$$

and the time-evolved operator $\hat{q}(t)$ is $e^{i\hat{H}t/\hbar} \hat{q} e^{-i\hat{H}t/\hbar}$.

Specifically, the RPMD approximation to Eq. (1.10) is

$$\tilde{C}_{qq}(t) = \frac{1}{Q_n} \int d^n q \int d^n v \, e^{-\beta H_n(q,v)} \bar{q}(0) \bar{q}(t) \qquad (1.12)$$

where $\bar{q}$ is the bead-averaged position

$$\bar{q}(t) = \frac{1}{n} \sum_{j=0}^{n-1} q_j(t) \, , \qquad (1.13)$$

and the pair $(q(t), v(t))$ are evolved by the RPMD equations of motion in Eq. (1.9) with initial conditions drawn from the classical Boltzmann-Gibbs measure.

The RPMD equations of motion can be compactly rewritten as

$$\begin{bmatrix} \dot{q} \\ \dot{v} \end{bmatrix} = A \begin{bmatrix} q \\ v \end{bmatrix} + \begin{bmatrix} 0 \\ F(q)/m_n \end{bmatrix} \, , \ \text{where } A = \begin{bmatrix} 0 & I \\ -\Omega^2 & 0 \end{bmatrix} \, , \qquad (1.14)$$

$F(q) = -\nabla V_n^{\text{ext}}(q)$, $I$ is an $n \times n$ identity matrix, $0$ is an array of zeros, and $\Omega^2$ is the $n \times n$ symmetric positive semi-definite matrix

$$\Omega^2 = \omega_n^2 \begin{bmatrix} 2 & -1 & 0 & \cdots & 0 & -1 \\ -1 & 2 & -1 & 0 & \cdots & 0 \\ & \ddots & \ddots & \ddots & & \\ & & \ddots & \ddots & \ddots & \\ 0 & \cdots & 0 & -1 & 2 & -1 \\ -1 & 0 & \cdots & 0 & -1 & 2 \end{bmatrix} . \qquad (1.15)$$

We recognize $\Omega^2$ as the 1D discrete Laplacian endowed with periodic boundary conditions; its spectral radius that scales as $n^2$, and since $\Omega^2$ is circulant, it can be diagonalized by the $n \times n$ orthogonal real discrete Fourier transform (DFT) matrix. In particular, the spectral decomposition of $\Omega$ can be written as

$$\Omega = U \Omega_d U^{\mathsf{T}}, \ \text{where } \Omega_d = \text{diag}(0, \omega_{1,n}, \ldots, \omega_{n-1,n}) \qquad (1.16)$$

is a diagonal matrix of eigenvalues given by

$$\omega_{j,n} = \begin{cases} 2\omega_n \sin\left(\frac{\pi j}{2n}\right) & \text{if } j \text{ is even} \, , \\ 2\omega_n \sin\left(\frac{\pi(j+1)}{2n}\right) & \text{else} \, , \end{cases} \qquad (1.17)$$

In nontrivial applications, the RPMD equations of motion in Eq. (1.14) cannot be solved analytically. It is then necessary to employ approximate numerical integration of the equations of motion. As we discuss next, designing good numerical integrators for Eq. (1.14) is complicated by the interplay between the time-evolution of the free ring-polymer (obtained by setting $F = 0$ in Eq. 1.14) and the contributions from the physical forces $F$.

**Cayley removes instabilities in a free ring-polymer mode**

RPMD is an example of highly oscillatory Hamiltonian dynamics.[40] To understand why numerical integration of such systems is tricky and why the Cayley modification is needed, it helps to consider the equations of motion for a particular normal mode of the free ring polymer with Matsubara frequency $\omega > 0$:

$$\begin{bmatrix} \dot{q} \\ \dot{v} \end{bmatrix} = A \begin{bmatrix} q \\ v \end{bmatrix} \quad \text{where} \quad A = \begin{bmatrix} 0 & 1 \\ -\omega^2 & 0 \end{bmatrix}, \tag{1.18}$$

which are also the equations of motion for a linear oscillator with natural frequency $\omega$. If $\omega$ is large, Eq. (1.18) is highly oscillatory. Solving Eq. (1.18) amounts to approximating the matrix exponential $\exp(\Delta t A)$ where $\Delta t$ is a timestep size. A good $2 \times 2$ matrix approximation $M_{\Delta t}$ should satisfy:

**(P1) Accuracy** $\|M_{\Delta t} - \exp(\Delta t A)\| = O(\Delta t^3)$.

**(P2) Strong Stability** For all $\omega > 0$, and for all $\Delta t$ smaller than some constant independent of $\omega$, $M_{\Delta t}$ is a strongly stable symplectic matrix.

**(P3) Time-Reversibility** For all $\omega > 0$ and $\Delta t > 0$, $M_{\Delta t}$ is reversible with respect to the velocity flip matrix $R = \begin{bmatrix} 1 & 0 \\ 0 & -1 \end{bmatrix}$, i.e., $R M_{\Delta t} R = M_{\Delta t}^{-1}$.

We briefly comment on each of these criteria for a good approximation. Property (P1) is a basic requirement that ensures second-order accuracy on finite-time intervals. Property (P3) is particularly useful for sampling from the stationary distribution, since a reversible map can be readily Metropolized[41–43], and since time-reversibility in a volume-preserving numerical integrator leads to a doubling of the accuracy order (see Propositions 5.2 and Theorem 6.2 of Ref.[43], respectively). Property (P2) is the most interesting. A symplectic matrix $S$ is stable if all powers of the matrix $S$ are bounded. A symplectic matrix $S$ is *strongly stable* if $S$ is stable and all sufficiently close symplectic matrices are also stable. In other words, $S$ is strongly stable if there exists an $\epsilon > 0$, such that all symplectic matrices $S^\epsilon$ that are within a distance $\epsilon$ away from $S$ are also stable. A sufficient condition for $S$ to be strongly stable is that the eigenvalues of $S$ are on the unit circle in the complex plane and are distinct; both the necessary and sufficient conditions for strong stability of symplectic matrices are known.[44]

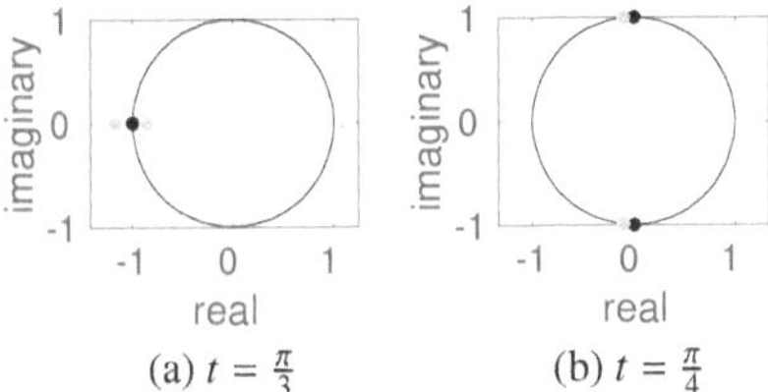



Figure 1.1: **Eigenvalues of $2 \times 2$ symplectic matrices.** Eigenvalues of a symplectic matrix $S = \exp(tA)$ (black dots) are plotted in the complex plane along with eigenvalues of a perturbed symplectic matrix $S^\epsilon = \exp((1/2)tB)\exp(tA)\exp((1/2)tB)$ (grey dots). The elements of $A$ and $B$ are specified in the text. For both values of $t$, $S$ is stable since its eigenvalues lie on the unit circle. When the eigenvalues of $S$ are not distinct, then as shown in (a), $S^\epsilon$ has an eigenvalue with modulus greater than one, and hence, $S^\epsilon$ loses stability. However, if the eigenvalues of $S$ are distinct, then $S$ is strongly stable, and as shown in (b), $S^\epsilon$ is stable since its eigenvalues remain on the unit circle.

Figure 1.1 illustrates the concept of strong stability. In particular, for different values of $t$ (as indicated in each panel), the black dots correspond to the eigenvalues of the symplectic matrix $S = \exp(tA)$ with $\omega = 3$, and the grey dots are the eigenvalues of a perturbation of $S$ which preserves the symplectic nature of the matrix, specifically

$$S^\epsilon = \exp((1/2)tB)\exp(tA)\exp((1/2)tB) \text{ where } B = \begin{bmatrix} 0 & \epsilon \\ \epsilon & 0 \end{bmatrix} \text{ and } \epsilon = 0.15. \text{ For}$$

any $t$, note that the two eigenvalues of $S$ are always on the unit circle, and hence, $S$ is always stable, but as the figure shows, $S$ is not always strongly stable. Indeed, in Figure 1.1 (a), we see that the two eigenvalues of $S$, represented by a single black dot, are both equal to $(-1, 0)$, which violates the condition for strong stability, and in this case, we see that one of the eigenvalues of $S^\epsilon$ has modulus greater than unity, which implies that $S^\epsilon$ is unstable. In Figure 1.1 (b), the two eigenvalues of $S$ are distinct and equal to $(0, \pm 1)$, and hence, $S$ is strongly stable. Since $S$ is strongly stable, and $\epsilon$ is sufficiently small, $S^\epsilon$ has eigenvalues that are on the unit circle, and hence, is itself stable. For a more detailed discussion of the concept of strong stability of symplectic matrices, see Section 42 of Ref.[45].

A natural candidate for an approximation $M_{\Delta t}$ that satisfies these criteria is the Verlet integrator, which is ubiquitous in the classical simulation of molecular systems.[46–49] For a single Matsubara frequency of the free ring polymer, the Verlet integrator gives

$$M_{\Delta t} = \begin{bmatrix} 1 - \frac{\Delta t^2 \omega^2}{2} & \Delta t \\ -\frac{1}{2}\Delta t \omega^2 (2 - \frac{\Delta t^2 \omega^2}{2}) & 1 - \frac{\Delta t^2 \omega^2}{2} \end{bmatrix}. \tag{1.19}$$

However, for $\Delta t > 2/\omega$, the eigenvalues of $M_{\Delta t}$ are real and distinct, so that one of them has modulus $> 1$, and therefore the powers of $M_{\Delta t}$ grow exponentially. Thus,

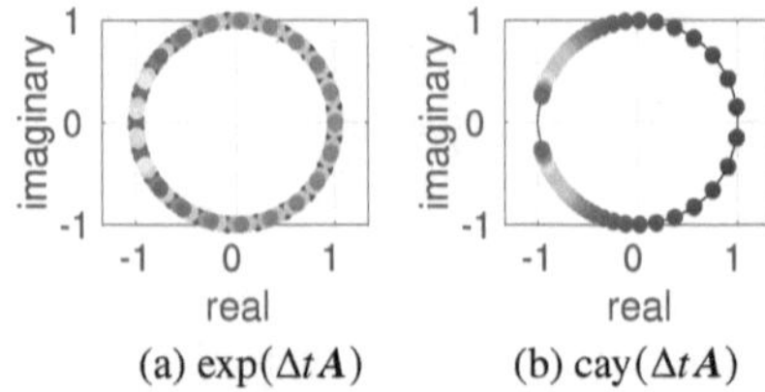

(a) exp($\Delta t A$)        (b) cay($\Delta t A$)

Figure 1.2: **Eigenvalues of the exponential vs. Cayley maps.** Eigenvalues of exp($\Delta t A$) (a) and cay($\Delta t A$) (b) at 50 different timestep sizes between 0.05 and 5.0 (evenly spaced) and with $\omega = 3$, color-coded from blue (smallest) through green and yellow to red (largest). For exp($\Delta t A$), the eigenvalues rotate around the unit circle multiple times. However, for cay($\Delta t A$), the eigenvalues start near $(1, 0)$, but never reach $(-1, 0)$. Since the eigenvalues of cay($\Delta t A$) are always distinct, it provides strong symplectic stability, whereas the matrix exponential loses strong stability every time the eigenvalues hit the horizontal axis. In both panels, the eigenvalue associated with the ring-polymer centroid motion is excluded.

numerical stability requires $\Delta t < 2/\omega$, and Verlet does not satisfy (P2), since this numerical stability requirement is not uniform in $\omega$.

Surprisingly, the exact solution for the normal-mode dynamics also does not satisfy (P2). To see why, note that the eigenvalues of the matrix exponential exp($\Delta t A$) are $e^{\pm i\omega\Delta t}$ and (P2) requires that $e^{i\omega\Delta t} \neq e^{-i\omega\Delta t}$ which is violated if and only if

$$\Delta t = \frac{\pi k}{\omega} \quad \text{for all } k \geq 1 . \tag{1.20}$$

At these timesteps, the exact solution violates strong stability. This is illustrated in Figure 1.2 (a), where the two eigenvalues of exp($\Delta t A$) are plotted in the complex plane for a range of time-step sizes. Although the two eigenvalues of exp($\Delta t A$) lie on the unit circle for all $\Delta t$, strong stability fails to hold whenever the eigenvalues are both equal to $(\pm 1, 0)$.

A simple strategy to avoid these artificial resonances is to use a random timestep size $\delta t$, e.g., take as timestep size an exponential random variable $\delta t$ with mean $\Delta t$. Averaging exp($\delta t A$) over the exponential probability density function yields $M_{\Delta t} = \mathbb{E}(\exp(\delta t A)) = (I - \Delta t A)^{-1}$, where here $I$ is the $2 \times 2$ identity matrix. Unfortunately, as can be easily verified, this matrix satisfies none of our criteria: it is neither symplectic, nor reversible, nor sufficiently accurate. However, we can easily turn this approximation into one that satisfies (P1), by simply composing $1/2$ step of this integrator with $1/2$ step of its adjoint $M_{\Delta t}^{-1}$. This correction yields the Cayley transform of the matrix $\Delta t A$,

$$\text{cay}(\Delta t A) \equiv (I - (1/2)\Delta t A)^{-1}(I + (1/2)\Delta t A). \tag{1.21}$$

In fact, the Cayley transform satisfies all three of the specified criteria for a good numerical integrator. It is time-reversible since $\boldsymbol{R}\,\mathrm{cay}(\Delta t A)\boldsymbol{R} = (\boldsymbol{R} - (1/2)\Delta t \boldsymbol{R}A)^{-1}(\boldsymbol{R} + (1/2)\Delta t A\boldsymbol{R}) = \mathrm{cay}(\Delta t A)^{-1}$, where we used that $\boldsymbol{R}^{-1} = \boldsymbol{R}$. It is a symplectic matrix since

$$\mathrm{cay}(\Delta t A)^T \boldsymbol{J}\,\mathrm{cay}(\Delta t A) = \boldsymbol{J} \quad \text{where} \quad \boldsymbol{J} = \begin{bmatrix} 0 & 1 \\ -1 & 0 \end{bmatrix}$$

where we used the fact that $A$ is a Hamiltonian matrix (See Ref.[50], Section 2.5). More importantly, it is a strongly stable symplectic matrix for all $\Delta t > 0$, as illustrated in Figure 1.2 (b); in contrast with the exponential map, for all $\omega > 0$ and $\Delta t > 0$ the eigenvalues of the Cayley map are $(4 - \Delta t^2\omega^2 \pm 4i\Delta t\omega)/(4 + \Delta t^2\omega^2)$, which are distinct and of unit modulus. Thus, not only is every matrix power of $\mathrm{cay}(\Delta t A)$ bounded, but the Cayley map is strongly stable uniformly in $\omega$ and $\Delta t$.

**Cayley removes instabilities in microcanonical RPMD**

For numerical integration of the conservative RPMD equations of motion (Eq. 1.9 or Eq. 1.14), it is standard practice[9,16] to employ a symmetrically split second-order integrator of the form in Eq. (1.2).

Furthermore, it is standard practice to exactly perform the free ring-polymer time evolution step,[16] using an exponential map of the form $\exp(\Delta t L_0) = \exp(\Delta t A)$ where $A$ is the matrix associated with the dynamics of the free ring-polymer Hamiltonian,

$$\begin{bmatrix} \dot{q} \\ \dot{v} \end{bmatrix} = A \begin{bmatrix} q \\ v \end{bmatrix}. \tag{1.22}$$

In practice, the exact exponential map is executed by successively *(i)* changing from the Cartesian bead positions and velocities to the normal modes of the free ring polymer, *(ii)* numerically integrating each of the uncoupled normal mode equations of motion, and *(iii)* translating the time-evolved normal mode coordinates back into the Cartesian bead positions and velocities. Therefore, the numerical stability of standard RPMD numerical integration may be analyzed in normal mode coordinates, where the free ring-polymer equations of motion in Eq. (1.22) decouple into a system of $n$ independent oscillators with natural frequencies given by the eigenvalues of the matrix $\boldsymbol{\Omega}$ in Eq. (1.17).

By applying Eq. (1.20) to each normal mode coordinate, we find that strong stability of the exact free ring-polymer time evolution is violated when

$$\Delta t = \frac{\pi k}{\omega_j} \quad \text{for all } k \geq 1 \text{ and } 1 \leq j \leq n - 1. \tag{1.23}$$

Unstable pairs of $\Delta t$ and $n$ are plotted using solid lines in Fig. 1.3(b) for selected values of $j$ and $k$. The horizontal asymptotes in this figure reflect the fact that the eigenvalues of $\boldsymbol{\Omega}$ converge to the eigenvalues of the continuous Laplacian endowed with periodic boundary conditions.

Unlike the exact free ring-polymer step used in standard RPMD numerical integration, the Cayley modification $\exp(\Delta t L_0) \approx \mathrm{cay}(\Delta t A)$ is strongly stable for all $\Delta t > 0$ uniformly in $n$. To see this, note that the Cayley transform can be equivalently computed in either bead or normal mode coordinates. More precisely, let $\Omega^2 = U\Omega_d^2 U^{\mathrm{T}}$ by the spectral decomposition of $\Omega$ given in Eq. (1.16). Direct computation then shows that

$$
\mathrm{cay}(\Delta t A) = \begin{bmatrix} U & 0 \\ 0 & U \end{bmatrix} \mathrm{cay}\left(\Delta t \begin{bmatrix} 0 & I \\ -\Omega_d^2 & 0 \end{bmatrix}\right) \begin{bmatrix} U^{\mathrm{T}} & 0 \\ 0 & U^{\mathrm{T}} \end{bmatrix} .
$$

Using this correspondence, one can invoke the preceding results on the one-dimensional oscillator, to conclude that $\mathrm{cay}(\Delta t A)$ is second-order accurate, strongly stable symplectic, and time-reversible.

Since the Cayley transform meets our criteria (P1)-(P3), and under suitable conditions on the force $F$, the Cayley modification to the RPMD numerical integrator is provably stable and second-order accurate on finite-time intervals with a stability requirement that is uniform with respect to the number of ring polymer beads. **On the other hand, standard RPMD integrators may display artificial resonance instabilities because the free RP step is not always strongly stable.** These instabilities often manifest as exponential growth in energy when strong stability is lost, as will be discussed in Section 1.4.

We emphasize that the improved numerical stability of the Cayley modification comes at zero cost in terms of algorithmic complexity or computational expense, and it preserves the same order of accuracy for the overall timestep. Use of this improved integration algorithm simply involves replacing the exact normal mode free ring-polymer step in the standard RPMD integrator with the Cayley modification.

**Algorithmic comparison: Standard vs. Cayley**

For complete clarity, we now present a side-by-side comparison of the full RPMD timestep (Eq. 1.2) with the free ring-polymer motion $\exp(\Delta t L_0)$ implemented using either the standard exponential map (i.e., exact normal mode evolution) or via the Cayley modification. In both cases, the full RPMD timestep associated with the splitting in Eq. (1.2) is implemented using the algorithm

$$
\begin{aligned}
&\textbf{Velocity half-step:} &&v \leftarrow v + \tfrac{\Delta t}{2}\tfrac{F}{m_n} \\
&\textbf{Free ring-polymer step:} &&(q, v) \leftarrow \mathrm{FRP}(q, v; \Delta t) \\
&\textbf{Force evaluation:} &&F = -\nabla V_n^{\mathrm{ext}}(q) \\
&\textbf{Velocity half-step:} &&v \leftarrow v + \tfrac{\Delta t}{2}\tfrac{F}{m_n}
\end{aligned}
\tag{1.24}
$$

In standard RPMD numerical integration, the free ring-polymer step is performed exactly, using:

1. Convert bead Cartesian coordinates to normal modes using the orthogonal transformation:

$$\varrho = Uq \qquad \text{and} \qquad \varphi = Uv \qquad (1.25)$$

where $U$ is the real DFT matrix defined in Eq. (1.16).

2. From $t$ to $t + \Delta t$, exactly evolve the free ring polymer in the normal mode coordinates:

$$\begin{pmatrix} \varrho_j(t + \Delta t) \\ \varphi_j(t + \Delta t) \end{pmatrix} = \exp(\Delta t A_j) \begin{pmatrix} \varrho_j(t) \\ \varphi_j(t) \end{pmatrix} \qquad (1.26)$$

where

$$A_j = \begin{bmatrix} 0 & 1 \\ -\omega_j^2 & 0 \end{bmatrix},$$

for $0 \le j \le n - 1$ with $\omega_j$ defined in Eq. (1.17).

3. Convert back to bead Cartesian coordinates using the inverse of $U$, which is just its transpose, since $U$ is orthogonal.

In the Cayley modification, the only change is to use the following in place of Eq. (1.26):

$$\begin{pmatrix} \varrho_j(t + \Delta t) \\ \varphi_j(t + \Delta t) \end{pmatrix} = \text{cay}(\Delta t A_j) \begin{pmatrix} \varrho_j(t) \\ \varphi_j(t) \end{pmatrix}, \qquad (1.27)$$

where cay is the Cayley transform given in Eq. (1.21).

## 1.3 Results for RPMD

In this section, we demonstrate the numerical integration of the microcanonical RPMD equations of motions (Eq. 1.14). Specifically, we compare the performance of the standard RPMD integrator, which involves exact integration of the free ring-polymer modes (Eq. 1.26) and our refinement in which the Cayley modification is used (Eq. 1.27). Results are presented for simple one-dimensional potentials, including

$$\text{Harmonic: } V(q) = \frac{1}{2}q^2 \qquad (1.28)$$

$$\text{Weakly anharmonic: } V(q) = \frac{1}{2}q^2 + \frac{1}{10}q^3 + \frac{1}{100}q^4 \qquad (1.29)$$

$$\text{Quartic: } V(q) = \frac{1}{4}q^4 \qquad (1.30)$$

and using a mass of $m = 1$.

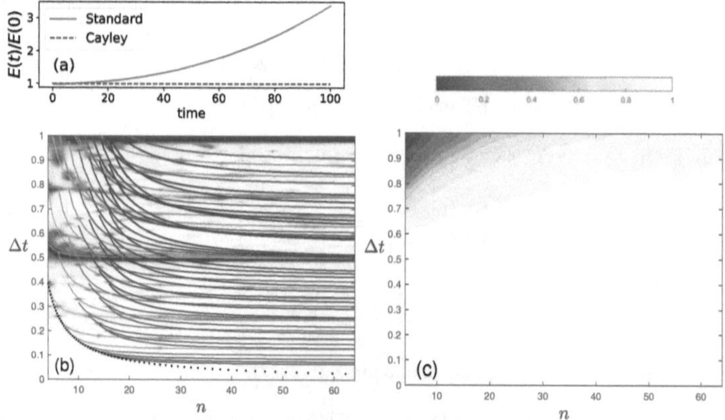

Figure 1.3: **Stability of RPMD trajectories on the harmonic oscillator potential.** (a) Representative trajectories performed using the standard RPMD integration scheme and using the Cayley modification. (b) Results for the standard RPMD numerical integration. The solid lines plot the instability condition in Eq. (1.23) for $k = \{1, \ldots, 10\}$ and $j = \{2, 4, \ldots, 16\}$. Higher values of $j$ are more blue, and higher values of $k$ are thicker. The dotted black line shows the maximum safe timestep defined in Eq. (1.31). The heatmap indicates the fraction of stable trajectories using standard RPMD integration. (c) The the fraction of stable trajectories using Cayley-modified RPMD integration. Results obtained at temperature $\beta = 1$.

We begin by numerically testing the conditions for loss of strong stability (Eq. 1.23) for the example of the harmonic potential (Eq. 1.28). Figure 1.3(a) shows a typical example of one of the approximately 25% of trajectories that fail for the standard RPMD integration scheme with $\beta = 1$, $n = 16$, and $\Delta t = 0.1$. The unstable trajectories start out with the typical values of ring-polymer energy in Eq. 1.6 (i.e., they are not the "hot" initial conditions from the tail of the thermal distribution), and they diverge to exponentially large energies after relatively short propagation time when run with the standard RPMD. All of these trajectories are stable when run with the Cayley modification.

The solid lines in Fig. 1.3(b) indicate predicted conditions for instability (Eq. 1.23). These analytical predictions are overlaid with a heatmap showing the fraction of stable RPMD trajectories on the harmonic potential using the standard RPMD integration scheme; for the purposes of the current section, a trajectories is deemed to be unstable if energy conservation associated with the ring-polymer Hamiltonian (Eq. 1.6) is violated by more than 10% within 100 time units of simulation. There are clear correlations in Fig. 1.3(b) between the predicted instabilities and observed simulation results.

Finally, Fig. 1.3(c) presents the corresponding heatmap for the Cayley-modified RPMD integration scheme. The Cayley modification preserves the conditions for strong stability, and the only numerically unstable trajectories are found for extremely large timesteps ($\Delta t > 0.6$). Comparison of Figs. 1.3(b) and (c) reveals the clear numerical advantages of the Cayley-modified RPMD integration scheme over the standard RPMD integration scheme.

Before proceeding, we emphasize the generality of the loss of strong stability with the standard RPMD numerical integrator: Eq. (1.23) makes no assumption with regard to the form of the physical potential, the dimensionality of the system, or the mass of the particles; it only depends on the temperature of the system and the number of ring-polymer beads in relation to the size of the integration timestep. Considering Eq. (1.23) for the $k = 1$ index and the highest Matsubara frequency of the ring-polymer, it is straightforward to show that the smallest possible timestep $\Delta t_*$ at which strong stability is violated is given by

$$\Delta t_* = \frac{\beta \hbar \pi}{2n} \, . \tag{1.31}$$

We thus arrive at a highly practical expression for the "maximum safe timestep" that depends only on $\beta$ and $n$, such that all smaller timesteps avoid the loss of strong stability associated with Eq. (1.23). In Fig. 1.3(b), this result is plotted (dotted, black line) and seen to follow the convex hull of smallest timesteps created by the other curves. In passing, we note that if $\beta$ corresponds to room temperature and $n = 64$, then the maximum safe timestep is 0.6 fs, which is strikingly consistent with the conventional 0.5 fs timestep employed in many PIMD simulations of liquid water.

Figure 1.4 confirms that the numerical instabilities of the standard RPMD integrator also manifest for anharmonic potentials. For both the weakly anharmonic (Eq. 1.29) and quartic (Eq. 1.30) potentials, we plot the fraction of stable trajectories as a function of timestep, comparing the standard RPMD integration scheme with the Cayley modification. Also shown are the fraction of stable classical mechanical trajectories (i.e., the 1-bead limit of RPMD) when integrated using the Verlet algorithm. Indeed, the standard RPMD integration scheme exhibits clear numerical instabilities at particular timesteps (which depend on the choice of $\beta$ and $n$), whereas the Cayley-modified integration scheme (like the classical integration scheme) avoids these pronounced instabilities. For the results in Fig. 1.4, the maximum safe timestep is $\Delta t_* \approx 0.029$. Note that the standard RPMD integration scheme on the weakly anharmonic potential does not exhibit significant loss of stability at this timestep, due to the fact that the unstable ring-polymer mode apparently does not sufficiently couple to the other modes on the timescale of the trajectories. However, the expected artifact at this timestep is indeed observed for the quartic potential. These results illustrate that the degree to which the resonance instabilities of standard RPMD integration manifest will depend on the application, but regardless of the system,

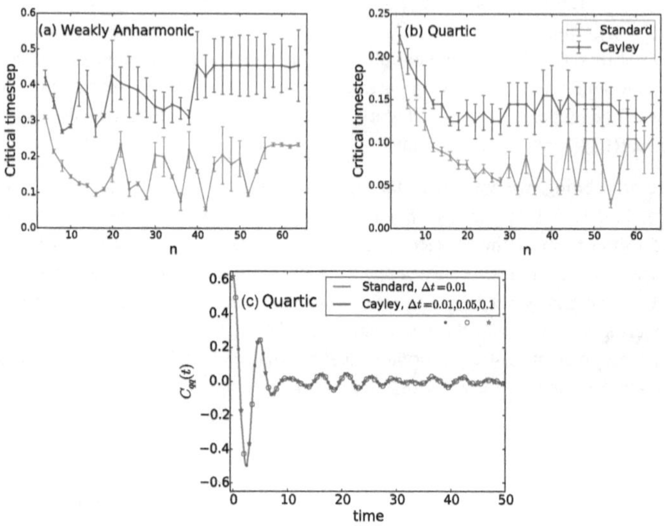


Figure 1.4: **Stability and Accuracy of RPMD trajectories on anharmonic potentials.** Percentage of stable RPMD trajectories using standard and Cayley-modified integration as a function of timestep, for the (a) weakly anharmonic and (b) quartic potentials. Results obtained using $n = 54$ and $\beta = 1$. Also included are classical MD results using the Verlet integrator. (c) For the quartic potential, comparison of the RPMD position time autocorrelation function obtained using standard integration with a small time-step where it is stable ($\Delta t = 0.01$) and using the Cayley modifiction with a range of larger timesteps ($\Delta t = 0.01$, filled circles; $\Delta t = 0.05$, empty circles; $\Delta = 0.10$, stars), indicating no significant loss of accuracy.

these resonance instabilities can be removed using the Cayley modification. Finally, panel (c) in this figure compares the accuracy of the standard and Cayley-modified RPMD integration schemes for the case of the quartic oscillator, revealing that even with time-steps that three-fold exceed the maximum safe timestep of the standard integration scheme, the Cayley-modified scheme shows negligible loss of accuracy in the trajectories.

Figure 1.5 explores the degree to which the Cayley modification enables the use of larger timesteps in comparison to the standard RPMD integration scheme. Defining the "critical timestep" as the largest value of $\Delta t$ for which 980 out of 1000 trajectories are stable, we compare this quantity for standard and Cayley-modified RPMD numerical integration as a function of the number of ring-polymer beads; the trends in the figure are insensitive to the precise definition of the critical timestep. Also

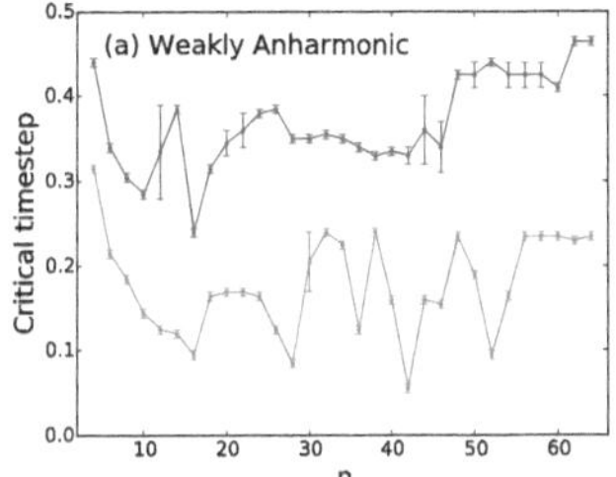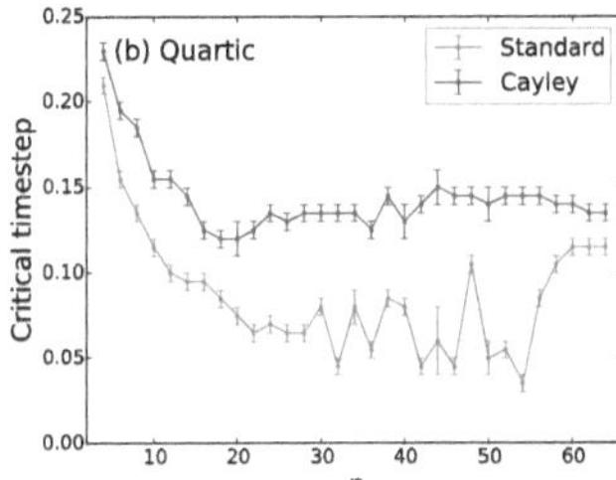



Figure 1.5: **Comparing largest stable timestep** as a function of the number of ring-polymer beads for the standard and Cayley-modified R PMD i ntegration s chemes o n the (a) weakly anharmonic and (b) quartic potentials. The critical timestep for the numerical simulations is defined i n t he t  ext. A lso s hown i s t he m aximum s afe t imestep f or the standard RPMD integration scheme (red dots). For classical MD integration using the Verlet algorithm, the critical timestep is 0.5 for the weakly anharmonic potential and 0.3 for the quartic potential. Results obtained at temperature $\beta = 1$.

shown is the maximum safe timestep for the standard RPMD integration scheme (Eq. 1.31). The improved stability of the Cayley-modified i ntegration s cheme is seen to consistently allow for the use of larger RPMD timesteps. The numerical behavior of the standard RPMD integration scheme closely tracks the predictions of the maximum safe timestep, although as seen previously, the resonance instabilities do not always manifest on the timescale of the simulated trajectories. Interestingly, for small $n$ in the quartic-oscillator simulations, the standard RPMD integration scheme actually underperforms the prediction of the maximum safe timestep, given that it exhibits large energy fluctuations ($> 10\%$) without fully encountering a resonance instability. In summary, using the maximum safe timestep for the standard RPMD integration scheme as a reference, the figure indicates that in these systems, the Cayley modification allows for substantial improvements in the allowed timestep size (three-fold or more for large $n$).

### 1.4   Results for T-RPMD

Thermostatted RPMD (T-RPMD) involves thermalization of the internal ring-polymer modes during RPMD dynamics, with the aims of improving sampling of the Boltzmann distribution[27] or avoiding the "spurious resonance" artifact that can appear in RPMD simulations of vibrational spectra.[11,29] Following Refs.[27] and[29], we implement T-RPMD using the splitting in Eq. (1.1), where $L_T$ corresponds to

$$\dot{v} = -\gamma v + \sqrt{2nm^{-1}\beta^{-1}}\gamma^{1/2}\dot{W}(t), \tag{1.32}$$

$\dot{W}(t)$ is a white-noise vector (since $W$ is an $n$-dimensional standard Brownian motion), and $\gamma$ is an $n \times n$ friction matrix defined such that $U^{\mathrm{T}}\gamma U$ is a diagonal matrix whose $k$th diagonal entry is equal to $\omega_k$ (Eq. 1.17). In normal mode coordinates (cf. Eq. 1.25), this thermostat is implemented by adding the following at the beginning and end of the full integration step outlined in Eq. 1.24:

$$\varphi_j(t + \Delta t) = e^{-\frac{\omega_j \Delta t}{2}} \varphi_j(t) + \sqrt{nm^{-1}\beta^{-1}}\sqrt{1 - e^{-\omega_j \Delta t}}\xi_j \, ,$$

where $\xi_j$ is a standard normal variate.

**Cayley removes non-ergodicity in T-RPMD**

Given that it helps to avoid spurious resonances,[29,30] one might expect that a Langevin thermostat can also eliminate the instabilities we have observed in standard RPMD integrators. This turns out to be only partly true. Here, we show that *(i)* lack of strong stability in the free RP step induces non-ergodicity in standard T-RPMD integrators, and *(ii)* the Cayley modification eliminates these non-ergodicity issues.

For this purpose, we revisit the simple case of a single free ring-polymer mode, as in Section 1.3. Consider Eq. (1.18) with a Langevin thermostat,

$$\begin{bmatrix} \dot{q} \\ \dot{v} \end{bmatrix} = K \begin{bmatrix} q \\ v \end{bmatrix} + \begin{bmatrix} 0 \\ \sqrt{2\beta^{-1}\gamma}\dot{W} \end{bmatrix} \, , \quad K = A + \begin{bmatrix} 0 & 0 \\ 0 & -\gamma \end{bmatrix} \, , \tag{1.33}$$

where $\gamma \geq 0$ is a friction factor and $\dot{W}(t)$ is a scalar white noise. The solution $(q(t), v(t))$ of Eq. (1.33) is a bivariate Gaussian with mean vector and covariance matrix given respectively by

$$\mu(t) = \exp(tK) \begin{bmatrix} q(0) \\ v(0) \end{bmatrix} \, ,$$

$$\Sigma(t) = 2\beta^{-1}\gamma \int_0^t \exp(sK) \begin{bmatrix} 0 & 0 \\ 0 & 1 \end{bmatrix} \exp(sK^{\mathrm{T}})ds \, . \tag{1.34}$$

In the limit as $t \rightarrow \infty$, the probability distribution of $(q(t), v(t))$ converges to the classical Boltzmann-Gibbs measure, which in this case, is a bivariate normal distribution with mean vector and covariance matrix given respectively by

$$\mu = \begin{bmatrix} 0 \\ 0 \end{bmatrix} \, , \quad \Sigma = \beta^{-1} \begin{bmatrix} \omega^{-2} & 0 \\ 0 & 1 \end{bmatrix} \, . \tag{1.35}$$

In this situation, the standard T-RPMD splitting in Eq. (1.1) inputs $(q_0, v_0)$ and outputs $(q_1, v_1)$ defined as

$$\begin{bmatrix} q_1 \\ v_1 \end{bmatrix} = OEO \begin{bmatrix} q_0 \\ v_0 \end{bmatrix} + \sqrt{\frac{1 - e^{-\gamma \Delta t}}{\beta}} \left( OE \begin{bmatrix} 0 \\ 1 \end{bmatrix} \xi_0 + \begin{bmatrix} 0 \\ 1 \end{bmatrix} \eta_0 \right) \tag{1.36}$$

where $\xi_0, \eta_0$ are independent standard normal random variables, $E = \exp(\Delta t A)$, and $O$ is the $2 \times 2$ matrix

$$O = \exp\left(\frac{\Delta t}{2}\Gamma\right) \, , \quad \Gamma = \begin{bmatrix} 0 & 0 \\ 0 & -\gamma \end{bmatrix} \, .$$

Moreover, the numerical solution after $N$ integration steps is a Gaussian vector with mean vector and covariance matrix given respectively by

$$\boldsymbol{\mu}_N = (\boldsymbol{OEO})^N \begin{bmatrix} q_0 \\ v_0 \end{bmatrix} , \quad \boldsymbol{\Sigma}_N = \sum_{j=0}^{N-1} (\boldsymbol{OEO})^j \boldsymbol{Q} (\boldsymbol{OE}^{\mathrm{T}}\boldsymbol{O})^j , \qquad (1.37)$$

where

$$\boldsymbol{Q} = \beta^{-1}(1 - e^{-\gamma \Delta t}) \left( \boldsymbol{OE} \begin{bmatrix} 0 & 0 \\ 0 & 1 \end{bmatrix} \boldsymbol{E}^{\mathrm{T}}\boldsymbol{O} + \begin{bmatrix} 0 & 0 \\ 0 & 1 \end{bmatrix} \right) .$$

From Eq. (1.20), if $\Delta t = k\pi/\omega$ for any $k \geq 1$, then $\boldsymbol{E}$ is not strongly stable. At these timesteps, the eigenvalues of the matrix $\boldsymbol{OEO}$ are given by $\lambda_+ = (-1)^k$ and $\lambda_- = (-1)^k \exp(-k\pi\gamma/\omega)$. By the Cayley-Hamilton theorem for $2 \times 2$ matrices,[51] we have the following representation of the $N$th power of $\boldsymbol{OEO}$

$$(\boldsymbol{OEO})^N = \frac{(\lambda_+)^N}{\lambda_+ - \lambda_-}(\boldsymbol{OEO} - \lambda_- \boldsymbol{I})$$
$$+ \frac{(\lambda_-)^N}{\lambda_- - \lambda_+}(\boldsymbol{OEO} - \lambda_+ \boldsymbol{I}) .$$

Since $|\lambda_+| = 1$, it follows from this representation that $\boldsymbol{\mu}_N$ does not converge to $\boldsymbol{\mu}$ in Eq. (1.35), since $\boldsymbol{\mu}_N$ clearly depends on the initial condition. Similarly, the covariance matrix $\boldsymbol{\Sigma}_N$ fails to converge to $\boldsymbol{\Sigma}$.

If we modify the above by replacing every instance of $\boldsymbol{E}$ with $\boldsymbol{C} = \mathrm{cay}(\Delta t \boldsymbol{A})$, the modified splitting is ergodic. More precisely, provided that the timestep is sufficiently small such that

$$2 > (1 + \cosh(\gamma \Delta t)) \left( \frac{4 - \Delta t^2 \omega^2}{4 + \Delta t^2 \omega^2} \right)^2 , \qquad (1.38)$$

then the eigenvalues of $\boldsymbol{OCO}$ are a complex conjugate pair with complex modulus $|\lambda_{\pm}| = \exp(-\gamma \Delta t/2)$. Hence, the matrix $\boldsymbol{OCO}$ is asymptotically stable. Under condition 1.38, the Cayley-modified scheme converges to the exact classical Boltzmann-Gibbs measure, in this example.

These results carry over to T-RPMD, where the free ring-polymer equations of motion in Eq. (1.22) decouple into a system of $n$ independent oscillators with natural frequencies given by the eigenvalues of the matrix $\boldsymbol{\Omega}$ in Eq. 1.17. Although the analysis of T-RPMD in this section was performed for the specific case of the splitting in Eq. (1.1) (i.e., the Bussi-Parrinello or OBABO splitting), we have confirmed that the same problem of non-ergodicity arises in the BAOAB splitting[52] and can likewise be fixed via the Cayley modification.

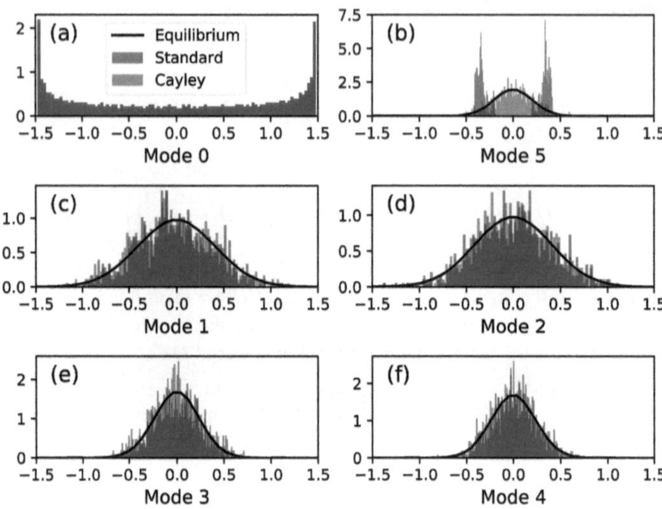

Figure 1.6: **Ergodicity of T-RPMD recovered with the Cayley modification, Example 1.** Normalized histograms of the ring-polymer normal mode displacement coordinates for a single trajectory (6 beads, $\beta = 1$), evolved on the harmonic potential with a timestep of $\Delta t = 0.26$. (a) The centroid mode, $\omega_j = 0$. (b) The predicted non-ergodic mode with $\omega_5 = 12$, (c-d), (e-f) pairs of modes with $\omega_1 = \omega_2 = 6$ and $\omega_3 = \omega_4 = 10.4$, respectively. Solid black line indicate the equilibrium distribution of the internal modes.

### T-RPMD numerical results

Figure 1.6 presents T-RPMD results on the harmonic potential (Eq. 1.28) using $n = 6$ and $\beta = 1$. For a single T-RPMD trajectory, we histogram the distribution of the normal mode coordinates that are sampled, employing the smallest timestep for which numerical instability is observed in the microcanonical case for this number of beads (see Fig. 1.3b); specifically, we use $\Delta t = 0.26$, which corresponds to the instability condition in Eq. (1.23) for the case of $n = 6$, $j = 5$, and $k = 1$. Using both standard and Cayley-modified T-RPMD integration, the trajectory is sampled at every timestep for a total of 770 timesteps.

The centroid mode (panel a) follows harmonic motion, that is decoupled from the other degrees of freedom. With both integrators, the lower-frequency ($j = 1 - 4$) internal ring-polymer modes are efficiently sampled and converge to the correct Gaussian distribution (panels c-f). However, the $j = 5$ mode behaves qualitatively differently, as predicted by Eq. (1.23), with the standard T-RPMD integrator showing clear non-ergodicity. The Cayley modification leads to ergodic sampling of all ring-

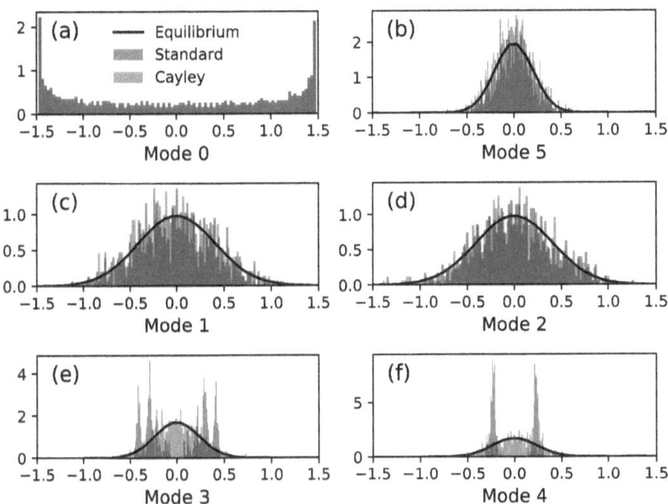

Figure 1.7: **Ergodicity of T-RPMD recovered with the Cayley modification, Example 2.** Normalized histograms of the ring-polymer normal mode displacement coordinates for a single trajectory (6 beads, $\beta = 1$), evolved on the harmonic potential with a timestep of $\Delta t = 0.3$. (a) The centroid mode, $\omega_j = 0$. (b) Unique highest frequency mode with $\omega_5 = 12$, (c-d) Modes with $\omega_1 = \omega_2 = 6$ (e-f) The predicted non-ergodic modes, $\omega_3 = \omega_4 = 10.4$. Solid black line indicate the equilibrium distribution of the internal modes.

polymer modes.

The lower frequency internal modes can also be afflicted with non-ergodicity at larger timesteps in this system. For the next-smallest unstable timestep in Fig. 1.3 ($\Delta t = 0.3$, which corresponds to the instability condition in Eq. (1.23) with $j = 3, 4$, and $k = 1$), the simulations were repeated. As predicted by the instability condition, modes 3 and 4 are found to be non-ergodic if sampled using the standard T-RPMD integrator (Fig. 1.7); again, ergodicity is recovered using the Cayley modification. The same non-ergodicity problems appear for anharmonic potentials using the standard T-RPMD integrator and can easily be avoided with use of the Cayley modification.

## 1.5  Summary

Strong stability is a relevant — and under-appreciated — concept for path-integral-based molecular dynamics methods. Without strong stability, numerical integration schemes are prone to numerical instabilities in the microcanonical case and non-ergodicity in the canonical case. Fortunately, one can easily imbue existing integration schemes, including those for PIMD, RPMD, T-RPMD, and many CMD methods, with strongly stability via the Cayley modification introduced here. This can be done without downside in terms of the computational cost, algorithmic complexity, or accuracy of the numerical integration scheme. The numerical results presented here suggest that this will have practical benefits for simulation studies, including improved stability, improved sampling efficiency, and improved efficiency via the use of larger MD timesteps.

While the Cayley transformation is familiar in the chemical physics literature in the context of the Crank-Nicolson propagator[53] for wavepacket dynamics,[54,55] and real-time path integrals[56] it has not to our knowledge been utilized for molecular dynamics, due to an under-appreciation of the property of strong stability.

The following Chapter describes how combination of the Cayley modification and the particular order of splitting can drastically improve the performance of the T-RPMD, relative to the standard scheme.[27,29]

# Chapter 2

# DIMENSION-FREE PATH-INTEGRAL MOLECULAR DYNAMICS

## 2.1 Introduction

Considerable effort has been dedicated to the development of numerical integration schemes for imaginary-time path-integral molecular dynamics (PIMD).[5] In comparison to standard classical molecular dynamics, PIMD numerical integration faces the additional challenge of the highly oscillatory dynamics of the ring-polymer internal modes. Work on PIMD numerical integration generally falls into two distinct categories. In the first, the PIMD equations of motion are *preconditioned* by modifying the ring-polymer mass matrix;[19–25,43,57] this approach, which includes the widely used staging algorithms,[31] causes the integrated trajectories to differ from those of the ring-polymer molecular dynamics (RPMD) model for real-time dynamics,[15,16] but it can lead to efficient[21–23] sampling of the quantum Boltzmann-Gibbs distribution.[2,3] In the second category, no modification is made to the ring-polymer mass matrix, i.e., the equations of motion are *non-preconditioned*.[9,16,27–30]

With the aim of providing useful models for real-time quantum dynamics, as well as simple and efficient al gorithms fo r eq uilibrium th ermal sa mpling, th e current work focuses on non-preconditioned PIMD numerical integration, notable examples of which include RPMD[15,16] and its thermostatted variant T-RPMD.[29] Numerical integration schemes for the latter methods typically employ symmetric factorizations of the time-evolution operator of the form[9,27–32,52,58,59]

$$e^{\Delta t \mathcal{L}} \approx e^{a\frac{\Delta t}{2}O} e^{\frac{\Delta t}{2}\mathcal{B}} e^{\frac{\Delta t}{2}\mathcal{A}} e^{(1-a)\Delta t O} e^{\frac{\Delta t}{2}\mathcal{A}} e^{\frac{\Delta t}{2}\mathcal{B}} e^{a\frac{\Delta t}{2}O} \tag{2.1}$$

where the operator $\mathcal{L} = \mathcal{A} + \mathcal{B} + O$ includes contributions from the purely harmonic free ring-polymer motion $\mathcal{A}$, the external potential $\mathcal{B}$, and a thermostat $O$. Note that the standard microcanonical RPMD numerical integration scheme is recovered in the limit of zero coupling to the thermostat, and that Eq. (2.1) yields the "OBABO" scheme of Bussi and Parrinello[32] when $a = 1$ and the "BAOAB" scheme of Leimkuhler[52] when $a = 0$. In Chapter 1 we emphasized that earlier PIMD numerical integration schemes had overlooked a fundamental aspect of the $\exp((\Delta t/2)\mathcal{A})$ sub-step of the time evolution in Eq. (2.1). Standard practice in these integration schemes has been to exactly evolve the harmonic free ring-polymer dynamics associated with $\exp((\Delta t/2)\mathcal{A})$ using the uncoupled free ring-polymer normal modes,[9,27,31] which was shown to lack the property of strong stability in the numerical integration, leading to resonance instabilities for microcanonical RPMD and loss of ergodicity for T-RPMD.[60] Use of the Cayley modification to the free ring-polymer motion was shown to impart strong stability to the time evolution, thereby improving numerical stability for microcanonical RPMD and restoring ergodicity for T-RPMD.[60]

In this Chapter, we focus on the accuracy of both statistical and dynamical properties of the OBABO and BAOAB schemes, as well as the corresponding integrators obtained when the exact free ring-polymer step is replaced by the strongly stable Cayley modification (OBCBO and BCOCB, respectively). Particular attention is

paid to the effect o f fi nite-timestep er ror wi th th ese in tegrators in th e li mit of
large bead numbers. Of these four integrators, it is found that only BCOCB is
"dimension-free," in the sense that the sampled ring-polymer position distribution
has non-zero overlap with the exact distribution in the infinite-bead l imit f or the
case of a harmonic potential. It is further shown that the OBCBO scheme can be
made dimension-free via the technique of force mollification. It i s s hown that the
newly introduced BCOCB integrator yields better accuracy than all other considered
non-preconditioned PIMD integrators and allows for substantially larger timesteps
in the calculation of both statistical and dynamical properties. Importantly, these
gains are made without loss of computational efficiency or algorithmic simplicity.

## 2.2  Non-preconditioned PIMD

Consider a one-dimensional molecular system with potential energy function $V(q)$
and mass $m$. The equations of motion for the corresponding $n$-bead ring polymer
held at constant temperature $T$ by a Langevin thermostat are

$$\dot{q}(t) = v(t) \, , \ \dot{v}(t) = -\mathbf{\Omega}^2 q(t) + \frac{1}{m_n}F(q(t))$$

$$- \mathbf{\Gamma}v(t) + \sqrt{\frac{2}{\beta m_n}}\mathbf{\Gamma}^{1/2}\dot{W}(t) \, . \tag{2.2}$$

Here, $W$ is an $n$-dimensional standard Brownian motion; $q(t) = (q_0(t), \ldots, q_{n-1}(t))$
is the vector of positions for the $n$ ring-polymer beads at time $t \geq 0$ and $v(t)$ are
the corresponding velocities; $m_n = m/n$ and $\beta = (k_BT)^{-1}$; and $F(q) = -\nabla V_n^{\text{ext}}(q)$,
where $V_n^{\text{ext}}$ is the contribution of the external potential defined in Eq. (1.7), $\mathbf{\Omega}^2$ is
the $n \times n$ symmetric positive semi-definite matrix defined in Eq. (1.15). Note that $\mathbf{\Omega}$
can be diagonalized by an $n \times n$ orthonormal real discrete Fourier transform matrix
$U$ (see Eq. 1.16). Finally, the matrix $\mathbf{\Gamma}$ in Eq. (2.2) is typically an $n \times n$ symmetric
positive semi-definite friction matrix of the form

$$\mathbf{\Gamma} = U \operatorname{diag}(0, \gamma_1, \ldots, \gamma_{n-1})U^{\mathrm{T}}, \tag{2.3}$$

where $\gamma_j$ is the friction factor in the $j$th normal mode.

In RPMD and T-RPMD calculations, one is often interested in the dynamics of
Eq. (2.2) with initial conditions drawn from the stationary distribution with non-
normalized density $\exp(-\beta H_n(q, v))$, where $H_n(q, v)$ is the ring-polymer Hamiltonian defined by

$$H_n(q, v) = H_n^0(q, v) + V_n^{\text{ext}}(q), \tag{2.4}$$

and $H_n^0(q, v) = (1/2)m_n \left(|v|^2 + q^{\mathrm{T}}\mathbf{\Omega}^2 q\right)$ is the free ring-polymer Hamiltonian.

The standard method for discretizing Eq. (2.2) is to use a symmetric splitting method
of the form of Eq. (2.1) that consists of a combination of three types of sub-steps:

(i) exact free ring-polymer evolution of timestep $\tau$,

$$\begin{pmatrix} q \\ v \end{pmatrix} \leftarrow \exp(\tau A) \begin{pmatrix} q \\ v \end{pmatrix}, \tag{2.5}$$

where $A = \begin{bmatrix} 0 & I \\ -\Omega^2 & 0 \end{bmatrix}$ is the Hamiltonian matrix associated to the free ring polymer,
(ii) velocity updates of timestep $\tau$ due to forces from the external potential,

$$v \leftarrow v + \tau \frac{1}{m_n} F(q), \tag{2.6}$$

and (iii) velocity updates of timestep $\tau$ due to the thermostat,

$$v \leftarrow \exp(-\tau \Gamma)v + \sqrt{\frac{1}{\beta m_n}} (I - \exp(-2\tau\Gamma))^{1/2}\xi, \tag{2.7}$$

where $I$ is the $n \times n$ identity matrix and $\xi$ is an $n$-dimensional vector whose components are independent, standard normal random variables. The acronyms OBABO and BAOAB indicate the order in which these sub-steps are applied, as indicated in Eq. (2.1) with $a = 1$ or $a = 0$, respectively.

In Chapter 1 we showed that the matrix exponential for the free ring-polymer evolution in Eq. (2.5) is not a strongly stable symplectic matrix, and as a consequence, the OBABO and BAOAB schemes can display non-ergodicity at timesteps $\Delta t = k\pi/\omega_{j,n}$ for any $1 \leq j \leq n$ and $k \geq 1$. We also identified a maximum safe timestep size $\Delta t_\star = \beta\hbar\pi/(2n)$, below which the matrix exponential is strongly stable. As $n \to \infty$, this maximum safe timestep goes to zero, such that no finite timestep for the scheme in Eq. (2.1) is safe in this limit from non-ergodicity.

This non-ergodicity motivates the Cayley modification[60] which consists of approximating the matrix exponential appearing in Eq. (2.5) with the Cayley transform. Specifically, for the Cayley-modified OBABO scheme (called OBCBO), we replace the exact free-ring polymer update of timestep $\tau = \Delta t$ with the Cayley transform given in Eq. (1.21)

For the Cayley-modified BAOAB scheme (called BCOCB), we replace the two exact free ring-polymer updates of half-timestep $\tau = \Delta t/2$ with $\mathrm{cay}(\Delta t A)^{1/2}$. While it might be expected that these half-timestep updates would instead be replaced with $\mathrm{cay}((\Delta t/2)A)$, such a choice leads to a loss of strong stability. Our use of the square root of the Cayley transform preserves strong stability, symplecticity, time reversibility, local third-order accuracy, and by definition satisfies $\mathrm{cay}(\Delta t A)^{1/2}\,\mathrm{cay}(\Delta t A)^{1/2} = \mathrm{cay}(\Delta t A)$. Furthermore, the square root of the Cayley transform is no more complicated to evaluate than the Cayley transform itself. Both the OBCBO and BCOCB Cayley modifications of Eq. (2.1) are ergodic for a fixed timestep, irrespective of the number of beads; moreover, like Eq. (2.1), the Cayley modified integrators exhibit locally third-order accuracy in the timestep and leave invariant the free ring-polymer Boltzmann-Gibbs distribution in the special case of a constant external potential ($V \equiv \text{const.}$).[60]

### 2.3 BCOCB avoids pathologies in the infinite bead limit

In this section, we show that of the OBABO, BAOAB, OBCBO, and BCOCB integration schemes, only BCOCB is dimension-free. Although the current section presents analytical results for the specific case of a harmonic external potential, these results are supported by numerical results for anharmonic external potentials in the subsequent sections.

To this end, consider the $j^{\text{th}}$ internal ring-polymer mode with frequency $\omega_{j,n}$, in the presence of a harmonic external potential $V(q) = (1/2)\Lambda q^2$ and a Langevin thermostat with friction $\gamma_j$. Expressed in terms of the normal mode coordinates $\varrho$ and $\varphi$, obtained from the Cartesian positions $q$ and velocities $v$ via Eq. (1.25), the non-preconditioned PIMD equations of motion for this mode are

$$
\begin{bmatrix} \dot{\varrho}_j(t) \\ \dot{\varphi}_j(t) \end{bmatrix} = K_j \begin{bmatrix} \varrho_j(t) \\ \varphi_j(t) \end{bmatrix} + \begin{bmatrix} 0 \\ \sqrt{2\beta^{-1}m_n^{-1}\gamma_j}\dot{W}_j(t) \end{bmatrix}
$$
$$
K_j = A_j + B + O_j,
$$
(2.8)

where $\dot{W}_j$ is a scalar white-noise, and we have introduced the following $2\times 2$ matrices

$$
A_j = \begin{bmatrix} 0 & 1 \\ -\omega_{j,n}^2 & 0 \end{bmatrix}, \quad B = \begin{bmatrix} 0 & 0 \\ -\Lambda/m & 0 \end{bmatrix}, \quad \text{and } O_j = \begin{bmatrix} 0 & 0 \\ 0 & -\gamma_j \end{bmatrix}.
$$

The solution $(\varrho_j(t), \varphi_j(t))$ of Eq. (2.8) is a bivariate Gaussian, and in the limit as $t \to \infty$, the probability distribution of $(\varrho_j(t), \varphi_j(t))$ converges to a centered bivariate normal distribution with covariance matrix

$$
\Sigma_j = \frac{1}{\beta m_n} \begin{bmatrix} s_j^2 & 0 \\ 0 & 1 \end{bmatrix}, \qquad s_j^2 = \frac{1}{\Lambda/m + \omega_{j,n}^2}.
$$
(2.9)

For this system, a single timestep of Eq. (1.1) can be compactly written as

$$
\begin{bmatrix} \varrho_j(t + \Delta t) \\ \varphi_j(t + \Delta t) \end{bmatrix} = M_j \begin{bmatrix} \varrho_j(t) \\ \varphi_j(t) \end{bmatrix} + R_j^{1/2} \begin{bmatrix} \xi_0 \\ \eta_0 \end{bmatrix},
$$
(2.10)

where $\xi_0$ and $\eta_0$ are independent standard normal random variables, and we have introduced the following $2 \times 2$ matrices

$$
M_j = e^{a\frac{\Delta t}{2}O_j} e^{\frac{\Delta t}{2}B} e^{\frac{\Delta t}{2}A_j} e^{(1-a)\Delta t O_j} e^{\frac{\Delta t}{2}A_j} e^{\frac{\Delta t}{2}B} e^{a\frac{\Delta t}{2}O_j}
$$

$$
R_j = \frac{1 - e^{-2(1-a)\gamma_j\Delta t}}{\beta m_n} N_j P N_j^{\mathrm{T}}
$$
$$
+ \frac{1 - e^{-a\gamma_j\Delta t}}{\beta m_n} \left( (M_j e^{-a\frac{\Delta t}{2}O_j})P(M_j e^{-a\frac{\Delta t}{2}O_j})^{\mathrm{T}} + P \right)
$$

where $\boldsymbol{P} = \begin{bmatrix} 0 & 0 \\ 0 & 1 \end{bmatrix}$ and $\boldsymbol{N}_j = e^{a\frac{\Delta t}{2}\boldsymbol{O}_j} e^{\frac{\Delta t}{2}\boldsymbol{B}} e^{\frac{\Delta t}{2}\boldsymbol{A}_j}$. The corresponding step for the Cayley modification is obtained by replacing $\exp((\Delta t/2)\boldsymbol{A}_j)$ in Eq. (2.10) with $\mathrm{cay}(\Delta t \boldsymbol{A}_j)^{1/2}$, which is given by

$$\mathrm{cay}(\Delta t \boldsymbol{A}_j)^{1/2} = \sqrt{\frac{1}{4 + \omega_{j,n}^2 \Delta t^2}} \begin{bmatrix} 2 & \Delta t \\ -\omega_{j,n}^2 \Delta t & 2 \end{bmatrix} . \tag{2.11}$$

A sufficient condition[a] for ergodicity of Eq. ) is

$$1 > \mathsf{A}_{j,\Delta t}^2 \cosh^2((\Delta t/2)\gamma_j) , \tag{2.12}$$

where

$$\mathsf{A}_{j,\Delta t} = \cos(\Delta t \omega_{j,n}) - \frac{(\Lambda/m)\Delta t}{2\omega_{j,n}} \sin(\Delta t \omega_{j,n}) .$$

For the Cayley modification of Eq. (2.10), Eq. (2.12) still provides a sufficient condition for ergodicity, except with

$$\mathsf{A}_{j,\Delta t} = -1 + \frac{8 - 2(\Lambda/m)\Delta t^2}{4 + \omega_{j,n}^2 \Delta t^2} .$$

Due to the lack of strong stability in the exact free ring-polymer evolution, Eq. (2.10) fails to meet the condition in Eq. (2.12) and becomes non-ergodic whenever $\Delta t = k\pi/\omega_{j,n}$ where $k \geq 1$ (see sec. 1.5 no such problem exists for the Cayley modification. Regardless, assuming that the condition in Eq. (2.12) holds, the numerical stationary distribution is a centered Gaussian with $2 \times 2$ covariance matrix $\boldsymbol{\Sigma}_{j,\Delta t}$ that satisfies the linear equation

$$\boldsymbol{\Sigma}_{j,\Delta t} = \boldsymbol{M}_j \boldsymbol{\Sigma}_{j,\Delta t} \boldsymbol{M}_j^{\mathrm{T}} + \boldsymbol{R}_j ,$$

for which the solution is

$$\boldsymbol{\Sigma}_{j,\Delta t} = \frac{1}{\beta m_n} \begin{bmatrix} s_{j,\Delta t}^2 & 0 \\ 0 & r_{j,\Delta t}^2 \end{bmatrix} \tag{2.13}$$

where the variance in the position and velocity marginal are $(\beta m_n)^{-1} s_{j,\Delta t}^2$ and $(\beta m_n)^{-1} r_{j,\Delta t}^2$ with

$$s_{j,\Delta t}^2 = \begin{cases} \dfrac{1}{\omega_{j,n}^2 + \frac{\Lambda \Delta t \omega_{j,n}}{m} \cot(\Delta t \omega_{j,n}) - (\frac{\Lambda \Delta t}{2m})^2} & a = 1 \\[2em] \dfrac{1}{\omega_{j,n}^2 + \frac{\Lambda \Delta t \omega_{j,n}}{2m} \cot(\frac{\Delta t}{2}\omega_{j,n})} & a = 0 \end{cases} \tag{2.14}$$

$$r_{j,\Delta t}^2 = \begin{cases} 1 & a = 1 \\[1em] \dfrac{2m\omega_{j,n} - \Lambda \Delta t \tan(\frac{\Delta t}{2}\omega_{j,n})}{2m\omega_{j,n}} & a = 0. \end{cases} \tag{2.15}$$

---

[a]In the special case when $\Lambda = 0$, the given condition for OBCBO corrects a sign error in Eq. 37 of Ref.[60].

For the Cayley modification of Eq. (2.10),

$$s_{j,\Delta t}^2 = \frac{4m}{4m - a\Delta t^2 \Lambda} s_j^2 , \tag{2.16}$$

$$r_{j,\Delta t}^2 = \frac{4m - (1-a)\Delta t^2 \Lambda}{4m} . \tag{2.17}$$

Note that these numerical stationary distributions are independent of the friction parameter $\gamma_j$, which is a benefit of schemes based on splitting the T-RPMD dynamics into Hamiltonian and thermostat parts, and using the exact Ornstein-Uhlenbeck flow in Eq. 2.7 to evolve the thermostat part. Moreover, comparing the exact covariance matrix in Eq. (2.9) with the finite-timestep approximations in Eqs. (2.13)-(2.17), note that in all cases $\Sigma_j = \lim_{\Delta t \to 0} \Sigma_{j,\Delta t}$. These results have previously been reported for the OBABO (Eqs. 2.14 and 2.15, $a = 1$) and BAOAB (Eqs. 2.14 and 2.15, $a = 0$) schemes[25,61] but not for the OBCBO (Eqs. 2.16 and 2.17, $a = 1$) or BCOCB (Eqs. 2.16 and 2.17, $a = 0$) schemes.

In the infinite bead limit, the exact and numerical position-marginals can be written as an infinite product of one-dimensional centered normal distributions with variances given by $(\beta m_n)^{-1} s_j^2$ and $(\beta m_n)^{-1} s_{j,\Delta t}^2$, respectively. By Kakutani's theorem,[62,63] these two distributions have a non-zero overlap if and only if the following series converges,

$$\sum_{j=1}^{\infty} \left(1 - \frac{s_j}{s_{j,\Delta t}}\right)^2 . \tag{2.18}$$

For OBABO and BAOAB, due to the oscillatory cotangent term appearing in $s_{j,\Delta t}$, the limit $\lim_{j \to \infty}(1 - s_j/s_{j,\Delta t})^2$ does not exist, and therefore, the series does not converge. For OBCBO, the $j$th summand of this series is

$$\frac{\Delta t^4 \Lambda^2}{16 m^2} \left(1 + \sqrt{\frac{4m - \Delta t^2 \Lambda}{4m}}\right)^{-2} ,$$

which more obviously leads to a divergent series. Therefore, for OBABO, OBCBO, and BAOAB, the numerical stationary distribution has no overlap with the exact stationary distribution in the infinite bead limit; it is in this sense that these schemes fail to exhibit the property of dimensionality freedom. Remarkably, BCOCB is exact in the position marginal and thus exhibits dimensionality freedom.

## 2.4 Consequences for the primitive kinetic energy expectation value

In the current section, we show that the non-overlap pathology of the OBABO, BAOAB, and OBCBO schemes causes a divergence with increasing bead number of the primitive path-integral kinetic-energy expectation value, an issue that is numerically well known for OBABO and BAOAB.[25,61,64,65] We further show that this divergence is fully eliminated via the BCOCB scheme — as expected.

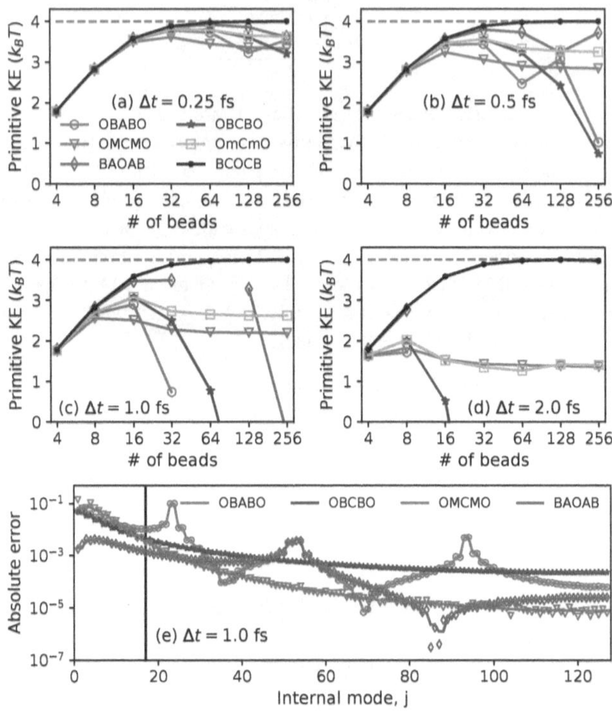


Figure 2.1: **Primitive kinetic energy expectation values** for a harmonic potential $V(q) = \frac{1}{2}\Lambda q^2$ with $\Lambda = 256$, $\hbar = m = 1$, and reciprocal temperature $\beta = 1$; choosing energies to be in units of $k_BT$ at room temperature (300 K), then $\beta\hbar \approx 25.5$ fs and $\Lambda = m\omega^2$ where $\omega = 3315$ cm$^{-1}$. (a-d) For various MD timesteps, the primitive kinetic energy expectation value as a function of the number of ring-polymer beads, with the exact kinetic energy indicated as a dashed gray line. The standard error of all visible data points in each plot is smaller than the symbol size. (e) Per-mode error in the variance of position coordinate of the normal modes for simulations run with 128 ring-polymer beads and a timestep of 1 fs; solid lines are analytic predictions from Eq. (2.24) with Eqs. (2.14) and (2.16) defining $s^2_{j,\Delta t}$ for the different schemes; points indicate the results of numerical PIMD simulations using the various integration schemes. The BCOCB scheme is not shown since it has zero error for all internal modes. The black vertical line indicates the crossover frequency ($\omega_x = 2/\Delta t$) for the error of OBCBO and OMCMO based on the bounds in Eqs. 2.35 and 2.36.

The primitive kinetic energy expectation value is given by[66,67]

$$\langle KE_{\text{prim}} \rangle = \frac{n}{2\beta} - \sum_{j=1}^{n} \frac{m_n \kappa_n^2}{2} \langle (q_j - q_{j-1})^2 \rangle \tag{2.19}$$

$$= \frac{1}{2\beta} + \sum_{j=1}^{n-1} \left( \frac{1}{2\beta} - \frac{m_n \omega_{j,n}^2}{2} \langle \varrho_j^2 \rangle \right) \tag{2.20}$$

where the first equality involves a sum over the ring-polymer beads in Cartesian coordinates (with $q_n = q_0$), and the second equality performs the summation in terms of the ring-polymer normal modes. The divergence of this expectation value is numerically illustrated for the simple case of a harmonic oscillator (Figs. 2.1a-d); note that for larger MD timesteps, the OBABO, BAOAB, and OBCBO schemes fail to reach a plateau with increasing bead number and dramatically deviate from the exact result (dashed line). The same divergence for OBABO and BAOAB has been numerically observed in many systems,[25,61,64,65] including liquid water which we discuss later. A striking observation from Figs. 2.1(a-d) is that the BCOCB exhibits no such divergence or error in the primitive kinetic energy expectation value at high bead number, regardless of the employed timestep.

Using Eq. (2.9), note that the contribution to the primitive kinetic energy expectation value from the $j$th ring-polymer mode is

$$\langle KE_j \rangle = \frac{1}{2\beta} \left( 1 - \omega_{j,n}^2 s_j^2 \right),$$

such that in the infinite-bead limit,

$$\lim_{n \to \infty} \sum_{j=0}^{n-1} \langle KE_j \rangle = \frac{\hbar}{4} \sqrt{\frac{\Lambda}{m}} \left( 1 + \frac{2}{e^{\hbar \beta \sqrt{\Lambda/m}} - 1} \right). \tag{2.21}$$

Similarly using Eq. (2.13), the $j$th-mode contribution to the kinetic energy from the finite-timestep numerical expectation value is

$$\langle KE_j \rangle_{\Delta t} = \frac{1}{2\beta} \left( 1 - \omega_{j,n}^2 s_{j,\Delta t}^2 \right). \tag{2.22}$$

Thus, the per-mode error in kinetic energy is

$$| \langle KE_j \rangle - \langle KE_j \rangle_{\Delta t} | = \frac{m_n \omega_{j,n}^2}{2} \rho_{j,\Delta t}, \tag{2.23}$$

where the per-mode error in the position marginal for internal mode $j$ is

$$\rho_{j,\Delta t} = \frac{1}{\beta m_n} \left| s_j^2 - s_{j,\Delta t}^2 \right|, \tag{2.24}$$

where $s_{j,\Delta t}$ is given by Eq. (2.14) for the cases of OBABO ($a = 1$) and BAOAB ($a = 0$) and by Eq. (2.16) for the cases of OBCBO ($a = 1$) and BCOCB ($a = 0$). Note that this error vanishes only for the BCOCB scheme, which satisfies $\rho_{j,\Delta t} = 0$ for each mode $j$, irrespective of the timestep $\Delta t$.

Eqs. (2.23) and (2.24) indicate that the primitive kinetic energy estimator is a sensitive measure of the finite-timestep error in the sampled ring-polymer position distribution associated with the high-frequency modes. Fig. 2.1(e) resolves this per-mode error, $\rho_{j,\Delta t}$, for each internal mode in simulations that employ a total of 128 beads, including results from OBABO (red), BAOAB (magenta) and OBCBO (blue) using a timestep of 1 fs, with the solid lines indicating the analytical predictions in Eq. (2.24) and with the dots indicating the result of numerical simulations. The analytical results are fully reproduced by the simulations. Note that the OBABO per-mode error exhibits dramatic spikes for $\omega_{j,n}\Delta t = k\pi$ where $1 \leq j \leq n$ and for some $k \geq 1$, which coincide with the loss of ergodicity of that integration scheme. The BAOAB scheme exhibits these resonance instabilities at even values of $k$. However, it is the failure of this per-mode error to sufficiently decay as a function of the mode number for all three of OBABO, BAOAB and OBCBO that gives rise upon summation to the divergence of the primitive kinetic energy expectation value, as seen for this particular timestep value in Fig. 2.1(d). Since $\omega_{j,n}^2 s_j^2 \to 1$ as $n \to \infty$, the convergence of $\sum_{j=1}^{\infty} |\langle KE_j \rangle - \langle KE_j \rangle_{\Delta t}|$ reduces to the convergence of the series $\sum_{j=1}^{\infty} \left| s_j^2 - s_{j,\Delta t}^2 \right|$, which diverges for both OBABO and OBCBO due to the same reasons as discussed in the section 2.3.

## 2.5 Dimensionality freedom for OBCBO via force mollification

The previous sections have demonstrated that whereas the BCOCB integrator exhibits dimensionality freedom, the OBCBO integrator does not. In the current section, we show that this shortcoming of OBCBO can be addressed by the use of force mollification, in which the external potential energy in Eq. (1.7) is replaced by

$$\tilde{V}_n^{\text{ext}}(\boldsymbol{q}) = V_n^{\text{ext}}(\text{sinc}(\tilde{\boldsymbol{\Omega}}\Delta t/2)\boldsymbol{q}), \tag{2.25}$$

where $\tilde{\boldsymbol{\Omega}}$ is any positive semi-definite $n \times n$ matrix that has the same eigenvectors as $\boldsymbol{\Omega}$ (Eq. 1.16) while possibly having different eigenvalues. Force mollification has not previously been employed for PIMD, although the strategy originates from a variation-of-constants formulation of the solution to Eq. (2.2);[68–71] specifically, the protocol in Eq. (2.25) is a generalization of the mollified impulse method.[68]

Use of force mollification in the current work can be motivated on physical grounds: In the absence of a physical potential, four of the considered integration schemes (OBABO, BAOAB, OBCBO, and BCOCB) leave invariant the exact free ring-polymer Boltzmann-Gibbs distribution. Therefore, the loss of any overlap between the exact stationary distribution of the position marginals in the infinite-bead limit for OBABO, BAOAB, and OBCBO must be attributed to the influence of the time evolution from the external potential in the schemes (i.e., the "B" sub-step) as implemented in Eq. 2.6; the BCOCB scheme does not suffer from this problem. To remove this pathology in the OBCBO scheme, we thus use mollification to taper down the external forces on the high-frequency modes, such that the resulting integration correctly reverts to free ring-polymer motion for those modes, which should become decoupled from the external potential as the frequency increases. The specific appearance of the $1/2$ factor in the sinc function argument ensures that the sinc function switches from its high-frequency effect to its low-frequency effect when the period of the Matsubara frequency is commensurate with $\Delta t$; the zero-frequency ring-polymer centroid mode is untouched by mollification.

Force mollification requires only a small algorithmic modification of the OBCBO integrator. Specifically, the "B" sub-step in Eq. 2.6 is replaced with

$$v \leftarrow v + \frac{\Delta t}{2} \frac{1}{m_n} \tilde{F}(q), \tag{2.26}$$

where the mollified forces are

$$\tilde{F}(q) = \mathrm{sinc}(\tilde{\Omega}\Delta t/2)F(\tilde{q}) = UD_{\Delta t}U^{\mathrm{T}}F(\tilde{q}) \tag{2.27}$$

where $\tilde{q} = UD_{\Delta t}U^{\mathrm{T}}q$ are the mollified bead positions, and where $D_{\Delta t}$ is the diagonal matrix of eigenvalues associated with $\mathrm{sinc}(\tilde{\Omega}\Delta t/2)$, i.e.,

$$D_{\Delta t} = \mathrm{diag}(\mathrm{sinc}(\tilde{\omega}_{0,n}\Delta t/2), \ldots, \mathrm{sinc}(\tilde{\omega}_{n-1,n}\Delta t/2)) \tag{2.28}$$

where $\tilde{\omega}_{j,n}$ is the $j$th eigenvalue of $\tilde{\Omega}$. In practice, the mollified forces are computed in normal mode coordinates as follows:

(a) Starting with the ring-polymer bead position in normal mode coordinates, obtain a copy of the mollified bead positions via

$$\tilde{q} = UD_{\Delta t}\varrho \ . \tag{2.29}$$

(b) Evaluate the external forces at the mollified ring-polymer bead positions, $F(\tilde{q})$.

(c) Apply the remaining mollification to the forces in Eq. 2.27 via

$$U^{\mathrm{T}}\tilde{F}(q) = D_{\Delta t}U^{\mathrm{T}}F(\tilde{q}) \ . \tag{2.30}$$

We emphasize that in comparison to the standard force update (Eq. 2.6) the use of the mollified force update (Eq. 2.26) introduces neither additional evaluations of the external forces nor $n \times n$ matrix multiplies associated with the discrete Fourier transform; it therefore avoids any significant additional computational cost.

This mollification scheme preserves reversibility and symplecticity as well as local-third order accuracy of the OBCBO scheme with timestep. We emphasize that the sinc-function-based mollification scheme in Eq. 2.26 is not unique, and alternatives can certainly be devised. Even within the functional form of the mollification in Eq. 2.26, flexibility remains with regard to the choice of the matrix $\tilde{\boldsymbol{\Omega}}$, which allows for mode-specificity in the way the mollification is applied. A simple choice for this matrix is $\tilde{\boldsymbol{\Omega}} = \boldsymbol{\Omega}$, such that mollification is applied to all of the non-zero ring-polymer internal modes. With this choice, we arrive at a fully-specified integration scheme that replaces the original "B" sub-step in Eq. 2.6 with the mollified force sub-step in Eq. 2.26; we shall refer to this force-mollified version of OBCBO integration scheme as "OMCMO." In the following sub-section, we propose a partially mollified choice for $\tilde{\boldsymbol{\Omega}}$ that further improves the accuracy.

For the harmonic external potential, all of the previously derived relations for OBCBO (most notably Eqs. 2.12, 2.16-2.17, and 2.23-2.24) also hold for OMCMO with $\Lambda$ suitably replaced by $\tilde{\Lambda}_j = \mathrm{sinc}^2(\omega_{j,n}\Delta t/2)\Lambda$. Note that $\tilde{\Lambda}_j \leq \Lambda$, since $\mathrm{sinc}^2(x) \leq 1$ for all $x \geq 0$, making clear that the mollification reduces the effect of the external potential on the higher-frequency internal ring-polymer modes.

We now show that mollifying the forces in the B substep fixes the pathologies of OBCBO in the infinite-bead limit, by restoring overlap between the sampled and exact stationary distributions. To see this, note that the $j$th summand in Eq. (2.18) for OMCMO satisfies

$$\left(1 - \frac{s_j}{s_{j,\Delta t}}\right)^2 \leq \left(1 - \frac{s_j^2}{s_{j,\Delta t}^2}\right)^2 \leq f(\omega_j\Delta t/2)\frac{\Delta t^4 \Lambda^2}{16m^2}$$

where $f(x) = ((1 - \mathrm{sinc}^2(x))/x^2 + \mathrm{sinc}^2(x))^2$, and we have used the infinite-bead limit for the ring-polymer internal-mode frequencies

$$\omega_j = \lim_{n\to\infty} \omega_{j,n} = \begin{cases} \dfrac{\pi j}{\hbar\beta} & \text{if } j \text{ is even}, \\ \dfrac{\pi(j+1)}{\hbar\beta} & \text{else}. \end{cases} \tag{2.31}$$

Using Eq. (2.31), we write

$$\sum_{j=1}^{\infty} f(\omega_j\Delta t/2) = 2\sum_{j=1}^{\lfloor\hbar\beta/(\pi\Delta t)\rfloor} f(j\pi\Delta t/(\hbar\beta)) + 2\sum_{j=\lceil\hbar\beta/(\pi\Delta t)\rceil}^{\infty} f(j\pi\Delta t/(\hbar\beta)).$$

Then the first term of the sum is bounded by

$$2\sum_{j=1}^{\lfloor\hbar\beta/(\pi\Delta t)\rfloor} f(j\pi\Delta t/(\hbar\beta)) \leq 2f(1)\hbar\beta/(\pi\Delta t) < 4\hbar\beta/(\pi\Delta t)$$

and for the second term we use

$$2 \sum_{j=\lceil \hbar\beta/(\pi\Delta t)\rceil}^{\infty} f(j\pi\Delta t/(\hbar\beta)) \leq F(1) + \hbar\beta/(\pi\Delta t) \int_1^{\infty} F(x)\,dx$$

where

$$F(x) = 2((1 - \mathrm{sinc}^2(x))/x^2 + 1/x^2)^2 \tag{2.32}$$

is monotone decreasing on $[1, \infty)$ with $F(1) \leq 4$ and $\int_1^{\infty} F(x)\,dx \leq 2$. We finally obtain

$$\sum_{j=1}^{\infty} f(\omega_j \Delta t/2) \leq 6\frac{\hbar\beta}{\pi\Delta t} + 4$$

$$\sum_{j=1}^{\infty} \left(1 - \frac{s_j}{s_{j,\Delta t}}\right)^2 \leq \left(6\frac{\hbar\beta}{\pi\Delta t} + 4\right)\frac{\Delta t^4 \Lambda^2}{16m^2}. \tag{2.33}$$

Again invoking Kakutani's theorem (Eq. 2.18), it follows that the numerical stationary distribution has an overlap with the exact stationary distribution. As a byproduct of this analysis, we can also quantify the amount of overlap between the exact and numerically sampled stationary distributions, [b] revealing that the total variation distance[72] between these distributions is given by

$$d_{\mathrm{TV}}(\mu, \mu_{\Delta t}) \leq \sqrt{\left(6\frac{\hbar\beta}{\pi\Delta t} + 4\right)\frac{\Delta t^2 \Lambda}{2m}}. \tag{2.34}$$

In summary, the force mollification strategy introduced here provably removes the pathologies due to the "B" sub-step in the case of a harmonic oscillator potential. Moreover, for any finite number of beads, the total variation distance between the exact and numerically sampled stationary distribution can be bounded by Eq. (2.34), and thus, OMCMO admits error bounds that are dimension-free.

Before proceeding, we first return to Fig. 2.1 to compare the accuracy of OMCMO with the un-mollified OBCBO scheme for the internal-mode position marginal of the harmonic oscillator. As seen in Fig. 2.1(e) for the results with a timestep of 1 fs, the per-mode error obtained by the mollified scheme (OMCMO, green) decays more rapidly with mode number than does OBCBO. Fig. 2.1(d) further illustrates that upon summation of the per-mode contributions, the OMCMO prediction for the primitive kinetic energy converges to a well-defined asymptote with respect to the number of ring-polymer beads, whereas OBCBO diverges as discussed earlier. Similar behavior is seen for shorter MD timesteps (panels a-c), although the failure of OBCBO becomes less severe with this range of bead numbers as the timestep is reduced.

---

[b]This quantification uses: (i) $d_{\mathrm{TV}} \leq 2d_{\mathrm{H}}$ where $d_{\mathrm{TV}}$ is the total variation distance and $d_{\mathrm{H}}$ is the Hellinger distance; and (ii) subadditivity of the squared Hellinger distance, which implies that $d_{\mathrm{H}}^2(\mu, \mu_{\Delta t}) \leq \sum_{j=1}^{\infty} d_{\mathrm{H}}^2(\mathcal{N}(0, s_j^2), \mathcal{N}(0, s_{j,\Delta t}^2)) \leq \sum_{j=1}^{\infty}(1 - s_j^2/s_{j,\Delta t}^2)^2 \leq (3\hbar\beta/(\pi\Delta t) + 2)\frac{\Delta t^4 \Lambda^2}{8m^2}$.

Although it is satisfying that mollification via OMCMO both formally and numerically ameliorates the problems of the OBCBO scheme in the high-bead-number limit, the OMCMO results in Fig. 2.1 are not ideal, since in some cases the OMCMO error is substantially larger than that of OBCBO when a modest number of beads is used (e.g., for 16 beads in panel d). This observation points to a simple and general refinement of the OMCMO scheme, which we discuss in the following subsection.

**Partial mollification**

Comparison of the per-mode errors from OBCBO and OMCMO in Fig. 2.1(e) reveals that lower errors for OMCMO are only enjoyed for internal modes that exceed a particular frequency (indicated by the vertical black line). This observation suggests that if a "crossover frequency" could be appropriately defined, then a refinement to OMCMO could be introduced for which mollification is applied only to the ring-polymer internal modes with frequency that exceed this crossover value.

For the case of a harmonic external potential, this crossover frequency $\omega_x$ can be found by comparing a bound for the per-mode error (Eq. 2.24) for OBCBO

$$\rho_{j,\Delta t} \leq \left( \frac{s_j^2}{m_n \omega_{j,n}^2 \beta} \frac{\Delta t^2 \Lambda}{4m - \Delta t^2 \Lambda} \right) \tag{2.35}$$

to that for OMCMO

$$\rho_{j,\Delta t} \leq g(\omega_{j,n}\Delta t/2) \left( \frac{s_j^2}{m_n \omega_{j,n}^2 \beta} \frac{\Delta t^2 \Lambda}{4m - \Delta t^2 \Lambda} \right), \tag{2.36}$$

where $g(x) = (1 - \text{sinc}^2(x))/x^2 + \text{sinc}^2(x)$. Since $g(x) \geq 1$ only when $x \leq 1$, we expect better accuracy if mollification is only applied to those ring-polymer internal modes with frequencies $\omega_{j,n} \geq \omega_x$, where $\omega_x = 2/\Delta t$. Although this result was derived for the case of a harmonic potential, it does not depend on $\Lambda$. We call this resulting partly mollified integration scheme "OmCmO." This scheme has the nice properties of OMCMO, including strong stability and dimensionality freedom.

Implementation of OmCmO is a trivial modification of OMCMO, requiring only that the diagonal elements of $\boldsymbol{D}_{\Delta t}$ in Eq. 2.28 are evaluated using

$$\text{sinc}(\tilde{\omega}_{j,n}\Delta t/2) = \begin{cases} 1 & \text{for } \omega_{j,n} < \omega_x \\ \text{sinc}(\omega_{j,n}\Delta t/2) & \text{otherwise,} \end{cases} \tag{2.37}$$

where $j = 0, \ldots, n - 1$. In physical terms, the emergence of $2/\Delta t$ in the crossover frequency is intuitive, since as was previously mentioned, it corresponds to having the ring-polymer mode undergo a full period per timestep $\Delta t$.

Finally, numerical results for the case of a harmonic potential (Figs. 2.1a-d) reveal that the partially modified OmCmO scheme (cyan) achieves both robust convergence of the primitive kinetic energy with increasing bead number, as well as better or comparable accuracy than the OBCBO and OMCMO integration schemes — as expected. However, it must be emphasized that for all panels of Fig. 2.1, the BCOCB scheme (which requires no force mollification) is by far the most accurate and stable.

## 2.6   Results for anharmonic oscillators

Having numerically characterized the performance of the various non-preconditioned PIMD integrators for the case of the harmonic oscillator external potential in Fig. 2.1, we now turn our attention to anharmonic external potentials. In this section, we consider both a weakly anharmonic (aHO) potential given by Eq. (1.29) and the more strongly anharmonic quartic potential given by Eq. (1.30).

All calculations are performed using $\hbar = 1$, $m = 1$, and $\beta = 1$. Assuming the system to be at room temperature (300 K), then the thermal timescale corresponds to $\beta\hbar \approx 25.5$ fs and $\Lambda = m\omega^2$, where $\omega = 3315$ cm$^{-1}$ for $\Lambda = 256$. The trajectories are performed with the centroid mode uncoupled from the thermostat (i.e., in the manner of T-RPMD); for the remaining $n - 1$ internal modes, simulations performed with the OBABO and BAOAB schemes use the standard[27,29] damping schedule of $\Gamma = \Omega$, and simulations performed using the Cayley modification (i.e., BCOCB, OBCBO, OMCMO, and OmCmO) use friction $\gamma_j = \min(\omega_{j,n}, 0.9\gamma_j^{\max}(\Lambda), 0.9\gamma_j^{\max}(0))$ for the $j^{\text{th}}$ mode, where $\gamma_j^{\max}(\Lambda)$ is the friction that saturates the inequality in Eq. (2.12); for the quartic potential, we set $\Lambda = 1$ in this calculation of $\gamma_j^{\max}$.

Panels (a) and (b) of Fig. 2.2 present kinetic energy expectation values for the aHO potential corresponding to $3315$ cm$^{-1}$ at room temperature. For the primitive kinetic energy expectation value, the results obtained using the various integration schemes with timesteps of both 0.5 fs (panel a) and 1.0 fs (panel b) are consistent with the observations for the harmonic potential in Fig. 2.1; specifically, the integrators without dimensionality freedom (OBABO, BAOAB, and OBCBO) fail to converge with increasing bead number, while the mollified integrators (OMCMO and OmCmO) smoothly converge with increasing bead number, and the partially mollified scheme (OmCmO) is consistently more accurate than OBCBO and OMCMO. However, it is also clear that BCOCB exhibits the best accuracy with increasing bead number, converging to the exact result without perceivable timestep error.

Figure 2.2(c-d) presents the corresponding results for the virial kinetic energy expectation value,

$$\langle KE_{\text{virial}} \rangle = \frac{1}{2\beta} - \frac{1}{2} \langle (\boldsymbol{q} - \bar{q}) \cdot \boldsymbol{F}(\boldsymbol{q}) \rangle \tag{2.38}$$

where $\bar{q}$ is the centroid (bead-averaged) position. Whereas the virial kinetic energy for all of the strongly stable integration schemes is well behaved, the OBABO and

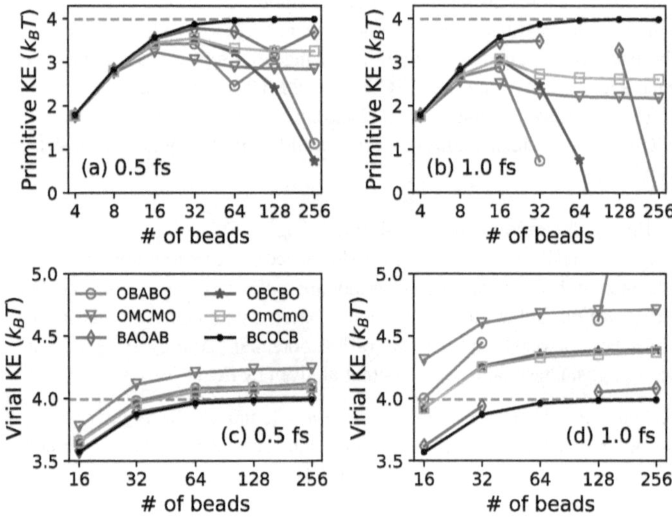

Figure 2.2: **Primitive and virial kinetic energy expectation values** as a function of bead number for the weakly anharmonic potential corresponding to 3315 cm$^{-1}$ at room temperature, with results obtained using a timestep of 0.5 fs (a,c) and 1.0 fs (b,d). The standard error of all visible data points in each plot is smaller than the symbol size. The exact kinetic energy is indicated with a dashed line.

BAOAB schemes perform erratically at large timesteps due to their provable non-ergodicities. (see sec. 1.5. Appealingly, the BCOCB scheme is consistently the most accurate for the virial kinetic energy expectation value, as it was for the primitive kinetic energy expectation value.

Figure 2.3(a-d) shows the results of the various numerical integration schemes for the primitive and virial kinetic energy expectation values, as a function of the MD timestep using 64 ring-polymer beads. Results are shown for both the aHO and the strongly anharmonic quartic oscillator. In all cases, the BCOCB scheme is consistently the most accurate across this array of model systems.

Finally, Fig. 2.3(e) illustrates the use of the BCOCB integrator for the calculation of real-time quantum dynamics via T-RPMD, replacing the often-employed OBABO integration scheme. Using 64 beads, the T-RPMD results are plotted for a range of integration timesteps. Strikingly, over the entire range of considered timesteps, BCOCB introduces negligible error in the calculated position time autocorrelation function; it is confirmed that these results are visually indistinguishable from those

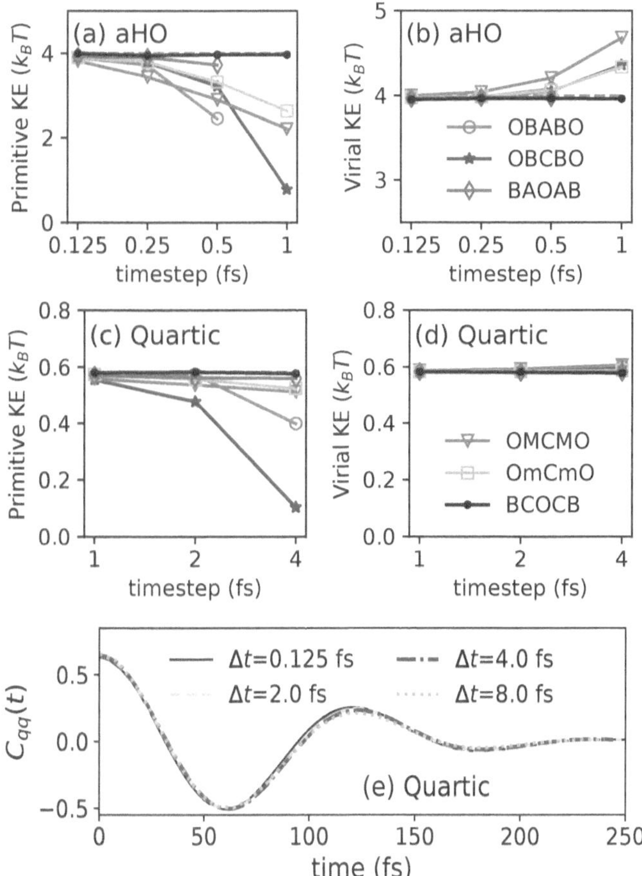


Figure 2.3: **Primitive and virial kinetic energy expectation values** as a function of the timestep for the weakly anharmonic potential corresponding to 3315 cm$^{-1}$ at room temperature (a,b), and the quartic potential (c,d). The exact kinetic energy is indicated with a dashed line. The standard error of all visible data points in each plot is smaller than the symbol size. Also, the position autocorrelation function (e) for the quartic oscillator at room temperature computed using T-RPMD with the BCOCB integrator. Results are obtained using 64 ring-polymer beads using timesteps of $\Delta t$ = 0.125, 2, 4, and 8 fs.

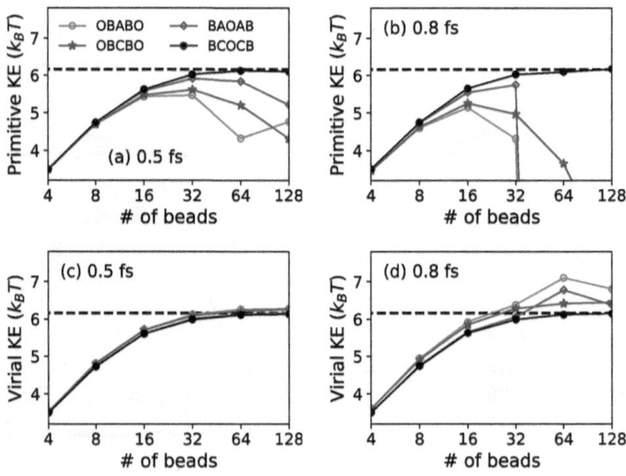


Figure 2.4: **Primitive and virial kinetic energy expectation values** as a function of the bead number per hydrogen atom in liquid water at 298 K and 0.998 g/cm$^3$ at timestep $\Delta t = 0.5$ fs (a, c) and $\Delta t = 0.8$ fs (b, d). The reference kinetic energy, obtained from a converged staging PIMD simulation at timestep $\Delta t = 0.1$ fs and bead number $n = 256$, is indicated with a dashed line. The standard error of all visible data points in each plot is smaller than the symbol size.

obtained using the OBABO integrator in the small-timestep limit.

## 2.7 Results for liquid water

The previous sections have demonstrated the strong performance of the BCOCB integrator for obtaining both PIMD statistics as well as real-time dynamics via the T-RPMD model, in model systems. Here, we test the accuracy and stability of the various un-mollified integration schemes (OBABO, OBCBO, BAOAB, and BCOCB) in liquid water, a high-dimensional and relatively complex system. Specifically, we consider a periodic 32-molecule water box at a temperature of 298 K and a density of 0.998 g/cm$^3$, as described by the q-TIP4P/F force field.[73]

In Fig. 2.4, we compare the accuracy achieved by the different integrators for the average kinetic energy per hydrogen atom as a function of the number of ring-polymer beads. As in previous sections, we consider both the primitive (Eq. (2.20)) and virial (Eq. (2.38)) estimators for the kinetic energy. For each choice of integrator, timestep, and bead number, the primitive and virial estimators for the kinetic energy of per hydrogen atom were averaged over a 1-nanosecond trajectory integrated in the manner of T-RPMD, i.e., with the centroid mode uncoupled from the thermostat; for the remaining $n - 1$ internal modes, simulations performed with the OBABO and BAOAB schemes use the standard[27,29] damping schedule of $\Gamma = \Omega$, and simulations performed using the Cayley modification use friction $\gamma_j = \min\{\omega_{j,n}, 0.9\gamma_j^{\max}(\omega_{\mathrm{OH}}^2), 0.9\gamma_j^{\max}(0)\}$, where $\gamma_j^{\max}(\Lambda/m)$ saturates the in-equality in Eq. (2.12) for the given values of $j$ and $\Lambda/m$ at the given time step, and $\omega_{\mathrm{OH}}$ is the OH-stretch frequency from the harmonic bending force field term in the q-TIP4P/F force field. Multi-nanosecond staging PIMD[25,31] simulations at a timestep of 0.1 fs were performed to obtain a bead-converged reference value for the H-atom kinetic energy, plotted as a dashed line in Figs. 2.4 and 2.5.

The primitive kinetic energy expectation values in panels (a) and (b) of Fig. 2.4 show similar trends to those seen in Figs. 2.1 and 2.2 for the harmonic and weakly anharmonic oscillators. For a 0.5-fs timestep (Fig. 2.4a), at which all integrators exhibit strong stability for ring polymers with up to 64 beads at the system temperature,[60] the OBABO, BAOAB, and OBCBO primitive kinetic energy estimates diverge from the converged result as the number of beads increases, in agreement with the proven result that the error in the ring-polymer configurational distribution generated with these schemes grows unboundedly with increasing bead number. At the larger, 0.8-fs timestep, (Fig. 2.4b), OBABO and BAOAB formally lose strong stability and their respective primitive kinetic energy estimates dramatically diverge for bead numbers greater than 32; the strongly stable OBCBO scheme also yields a divergent result for the same reason as in Fig. 2.4(a). As seen on the HO and aHO model systems, the primitive kinetic energy expectation value from the BCOCB integrator monotonically converges to the reference value with increasing bead number, avoiding any perceptible timestep error.

Fig. 2.4(c-d) shows the corresponding virial kinetic energy expectation values. For the smaller timestep of 0.5 fs, which is a common choice for path-integral simulations of water, all of the integrators perform similarly. However, upon increasing the timestep to 0.8 fs, significant differences in the performance of the integrators emerges, with only BCOCB avoiding perceptible timestep error.

To further compare the accuracy and stability of the OBABO, BAOAB, OBCBO, and BCOCB integrators, Fig. 2.5 considers the average kinetic energy per hydrogen atom obtained using 64 beads over a wide range of timesteps. These results show that BCOCB remains remarkably accurate for timesteps as large as 1.4 fs for liquid water, which corresponds to the limit of stability for Verlet integration of the centroid mode. In comparison, OBCBO diverges monotonically as the timestep increases, reaching unphysical values for the primitive expectation value and yielding sizable error (20%) for the virial expectation value. The erratic performance of both OBABO and BAOAB is due to the emergence of numerical resonance instabilities at timesteps greater than 0.6 fs at the employed bead number; indeed, the largest safe timestep at which OBABO and BAOAB remain strongly stable for $n = 64$, $\Delta t_\star \approx 0.63$ fs, precedes the range of timesteps in Fig. 2.5 for which these integrators

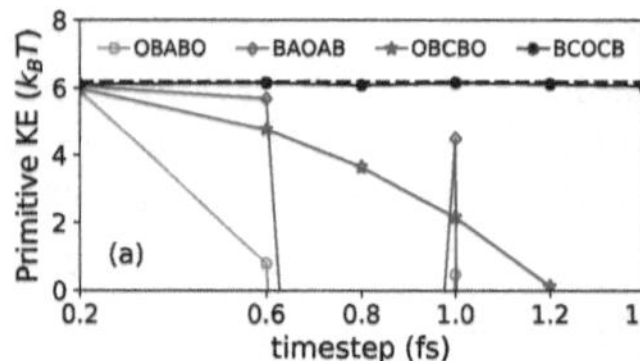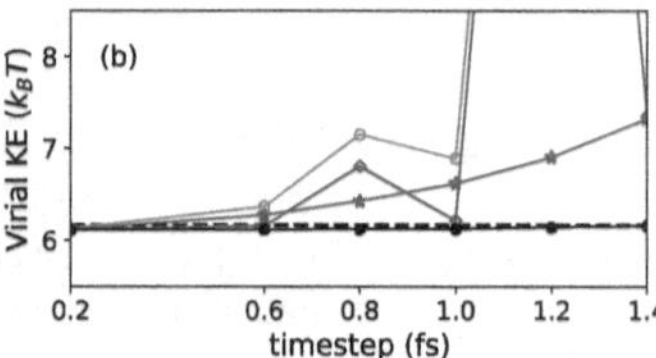

Figure 2.5: **Primitive and virial kinetic energy expectation values** as a function of the timestep per hydrogen atom in liquid water at 298 K and 0.998 g/cm$^3$, as described by a 64-bead ring polymer. The reference kinetic energy, obtained from a converged staging PIMD simulation at timestep $\Delta t = 0.1$ fs and bead number $n = 256$, is indicated with a dashed line. The standard error of all visible data points in each plot is smaller than the symbol size.

vary erratically.

Extending beyond statistics, we now consider the dynamical properties of liquid water. Given the superiority of the BCOCB scheme for the calculated statistical properties in Figs. 2.4 and 2.5, we present results that focus on this scheme in comparison to the most widely used OBABO scheme. In particular, we consider the liquid water infrared absorption spectrum,[11] which is proportional to $\omega^2 \tilde{I}(\omega)$ where the dipole spectrum $\tilde{I}(\omega) = \int_{\mathbb{R}} dt\, e^{-i\omega t} \tilde{C}_{\mu\cdot\mu}(t)$ is the Fourier transform of the Kubo-transformed dipole autocorrelation function $\tilde{C}_{\mu\cdot\mu}(t)$. The latter is approximated in the RPMD model by[9] $\tilde{C}_{\mu\cdot\mu}(t) = \frac{1}{N} \sum_{i=1}^{N} \langle \bar{\mu}_i(t) \cdot \bar{\mu}_i(0) \rangle$, where $N$ is the number of molecules in the liquid, $\bar{\mu}_i(t)$ is the bead-averaged dipole moment of molecule $i$ at time $t$, and the angle brackets denote averaging over the ring-polymer thermal distribution. To obtain the time-correlation functions and spectra shown in Fig. 2.6 for the OBABO and BCOCB integration schemes, 12-nanosecond T-RPMD trajectories were simulated for a ring polymer with 64 beads and timesteps ranging from 0.2 to 1.4 fs, using the same friction schedule as described for Figs. 2.4 and 2.5.

Along each trajectory, the velocities of all degrees of freedom in the system were drawn anew from the Maxwell-Boltzmann distribution every 20 picoseconds; the autocorrelation function was evaluated out to 2 picoseconds by averaging over staggered windows of that time-length within every 20-picosecond trajectory segment; and exponential-decay extrapolation was used to extend the autocorrelation function before evaluating its numerical Fourier transform to obtain the infrared absorption spectrum.

Fig. 2.6(a-b) present the dipole autocorrelation functions obtained using the OBABO and BCOCB integrators with a range of timesteps. For the OBABO integrator, the calculated correlation function is qualitatively incorrect for timesteps as large as 0.8 fs. For the BCOCB integrator, the resulting correlations functions are far more robust with respect to timestep. Although modest differences are seen in the exponential tail of the correlation function, the dynamics on vibrational timescales (see inset) is largely unchanged as the timestep is varied from 0.2 fs to 1.4 fs. Fig. 2.6(c) further emphasizes this point by showing the absorption spectrum that is obtained from the BCOCB time-correlation functions with the various timesteps. To minimize bias, we avoided any smoothing of the spectra shown in panel c. It is clearly seen that the librational and bending features (below 2500 $cm^{-1}$) are visually indistinguishable over the entire range of considered timesteps. To clarify the comparison for the stretching region above 3000 $cm^{-1}$, we smooth the raw spectra in that region by convolution against a Gaussian kernel with a width of 150 $cm^{-1}$ (see inset). Again, the robustness of the simulated spectrum over this span of timesteps is excellent, with the only significant effect due to finite-timestep error being a slight blue-shifting of the OH stretching frequency for the results using a 1.4-fs timestep, which is nearly three times larger than the typical value employed for the OBABO scheme for simulations with 64 beads. Taken together, these results indicate that the BCOCB integrator provides an excellent description of both PIMD statistics and T-RPMD dynamics in realistic molecular systems, substantially improving the accuracy and stability of previously employed numerical integrators.

## 2.8  Summary

In Chapter 1, we showed that essentially all schemes for the non-preconditioned equations of motion of PIMD, including the widely used OBABO scheme, lack strong stability due to the use of exact free ring-polymer time evolution in the "A" sub-step, and we proved that this lack of strong stability gives rise to a lack of ergodicity in the thermostatted trajectories. We further showed that ergodicity can be restored by simply replacing the "A" sub-step with the Cayley transform.

Here we show that a completely distinct — yet equally important — pathology exists in the "B" sub-step of previously developed non-preconditioned PIMD integrators, due to the outsized effect of the external potential on the dynamics of the high-frequency ring-polymer modes. Specifically, we show that previous integrators (including OBABO, BAOAB, and OBCBO) yield a numerical stationary distribution for which the overlap with the exact stationary distribution vanishes in the infinite-bead limit. We then show that this pathology is completely avoided in the BCOCB scheme, and we further show that the pathology can be eliminated for the OBCBO scheme by suitably mollifying the "B" sub-step, yielding the dimension-free non-preconditioned PIMD integrators, namely BCOCB, OMCMO, and OmCmO. Implementation of the dimension-free integration schemes involves no significant additional computational cost, no additional parameters, and no increase in algorithmic complexity in comparison to either OBABO or BAOAB. Furthermore, since the integrators considered here are all non-preconditioned, they can immediately be used for computing the equilibrium statistical properties as well as dynamical properties via the RPMD model. The numerical performance of the

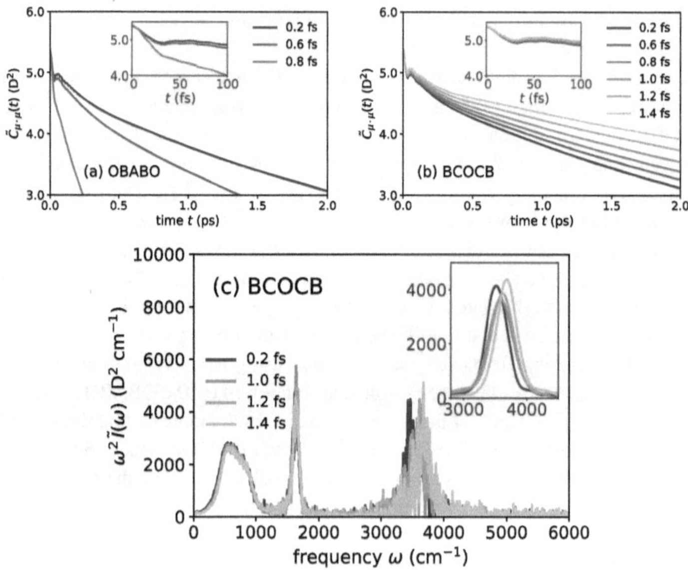

Figure 2.6: **Dynamical properties of liquid water** computed using T-RPMD with the (a) OBABO and (b,c) BCOCB integration schemes. Panels (a) and (b) present the Kubo-transformed dipole autocorrelation function computed with various timesteps, and panel (c) presents the absorption spectrum from the BCOCB correlation function at each timestep. The inset to panel (c) presents the OH stretching region with smoothing.

BCOCB scheme is particularly striking, yielding results that are markedly better in terms of accuracy and timestep stability than any of the other considered integrators. For liquid water, it is shown that BCOCB allows for timesteps as large as 1.4 fs while exhibiting minimal timestep error in the calculation of both equilibrium expectation values and the dipole absorption spectrum.

# Chapter 3

# APPLICATION OF PATH INTEGRALS TO HEAVY ISOTOPE EQUILIBRIA

## 3.1 Introduction

Since the development of BCOCB integrator (Chapter 2), it has been incorporated into the python-based *i-pi* wrapper package[74], Julia-based *NQCDynamics*[75], high-performance Tinker-HP[76] and used by others to study nuclear quantum effects on thermal conductivity,[77] reaction rates,[78,79] and electron transfer.[80]

The follow-up study by Rosa-Raices et al..[81] evaluated the integrators based on their respective rate of convergence to equilibrium and their efficiency at evaluating equilibrium expectation values. They considered a broader class of T-RPMD numerical integrators that exhibit strong stability and dimensionality freedom, just as the BCOCB integrator does and concluded that BCOCB excels over other known integrators in terms of accuracy, efficiency, and stability with respect to time step size based on both analytical and numerical results.[81]

We therefore turn our attention to the application of path-integral methods. Specifically, in the following Chapters we will study the heavy stable isotope equilibria between small gaseous molecules with PIMC. These effects are of great interest in geochemistry, as they can be used to identify (or at least constrain) the origin and history of a given sample. PIMC is especially suited for the study of the equilibration of heavy isotopes among isotopologues as (i) the effect of interest is purely statistical in nature; (ii) the deviation from randomness is due to the nuclear quantum effects; and (iii) the systems of interest include few physical dimensions (up to 11 atoms, i.e., up to 27 physical degrees of freedom).

## 3.2 Nomenclature of stable isotope equilibria

We begin by providing the nomenclature used to describe relative differences in the isotopic composition of phases and sites in the same molecule as well as clumped-isotope compositions. Although much of the notation used is standard in geochemistry, we have found that notational differences do exist between chemical and geochemical studies that can lead to confusion in regard to the meaning of what is actually being calculated or measured (as elaborated below). As such, we provide a complete explanation of the various nomenclature used here and note where prior work has used different notations.

Here we will denote stable heavy isotopes following the usual convention, i.e., by prepending the mass number of the isotope as a superscript to the atomic symbol, e.g., $^{13}C$ for carbon-13 with the exception of deuterium (hydrogen-2), which has its own symbol $^2$H $\equiv$ D. We will omit the mass number of the most abundant isotope of each chemical element, i.e., protium will be denoted as H, carbon-12 as C, oxygen-16 as O etc. to avoid using too many superscripts and more importantly to draw reader's attention to heavy isotope substitutions of a particular isotopologue.

**Isotopic differences between phases or species and between sites in the same molecule**

The "bulk" (i.e., average) isotopic composition of a given phase, species, or site within a molecule is traditionally given using the R notation:

$$^i R = \frac{[^{\text{heavy}} A]}{[^{\text{light}} A]} \tag{3.1}$$

where the superscript $i$ denotes the rare isotope of atom $A$ being examined. As an example $^{13}R = [^{13}\text{C}]/[\text{C}]$. Importantly, both rare and abundant isotopes are spread across all isotopologues in a system. We emphasize this because, as will be discussed below, relating equilibrium calculations to measured $R$ values requires additional considerations for any system that is not at a random distribution of isotopes amongst all isotopologues. The bulk isotopic compositions of molecules are generally reported using delta notation where:

$$\delta^i A_M = 1000 \times \left( \frac{^i R_M}{R_{\text{standard}}} - 1 \right). \tag{3.2}$$

The "standard" is the reference standard that all samples are reported relative to for that isotopic system. Here, the $^i A$ refers to the isotopic system being examined (e.g., $^{13}$C, $^{18}$O, or D). The "M" corresponds to the phase, species, or site. As an example, stable carbon isotopic composition of methane would be given as $\delta^{13}$C relative to the VPDB reference.[82]

Differences in the isotopic composition between sites within a molecule, phases, or species are given using alpha notation:

$$^i \alpha_{M-N} = \frac{^i R_M}{^i R_N} = \frac{\delta^i A_M + 1000}{\delta^i A_N + 1000} \tag{3.3}$$

As in the definition for delta notation (Eq. 3.2), in Equation (3.3), the M and N can represent two different phases (e.g, liquid water vs water vapor), species in a given phase (e.g., carbon dioxide and methane in gas phase for carbon isotopes), two sites within the same molecules (e.g., terminal methyl vs. center methylene carbon of propane), or the site of one molecule (e.g., the terminal methyl of propane) vs. the bulk isotope composition or site of another molecule in the same or different phase. The $\alpha$ is generally termed a "fractionation factor."

We note that our definition of isotopic differences between phases or species is the textbook definition for isotope fractionation factors.[83,84] However, there are two definitions used for site-specific isotope effects (also termed position-specific isotope effects and we use those terms interchangeably here). In one case, position-specific effects are given as difference between the isotopic composition at a given site vs. the bulk isotopic composition of the whole molecule. In the other case, the isotopic difference is given as the difference between two sites. These differences are both commonly represented using $\Delta$ notation (which we reserve below for clumped isotopes only) in which the isotopic difference in capital delta notation between the sites or a site and the bulk molecule is given. Indeed, both definitions are given in the recent review.[85] Which one is used often depends on the history of prior measurements or calculations for a given molecule type and/or what can be measured.

Here we report the site-specific calculations as isotopic differences between the sites. This is sometimes termed a "site preference." If a difference relative to an average is desired, it can be directly calculated from the difference in isotopic composition of the sites. We have reviewed the site/position specific information in detail because this difference in nomenclature has led to some confusion in terms of comparing theoretical and measurement-based methods of site-specific isotope effects in propane.[86–88] This is discussed in detail in section 6.5.

**Clumped isotope effects**

Clumped isotope effects describe the excess or deficit concentration of isotopologues with multiple rare isotopic substitutions relative to what would be expected given a random distribution of isotopes amongst all isotopologues.[89] Here we only report clumped isotope effects of doubly substituted species. We give the concentration of the rare (at natural abundance), doubly substituted species relative to that for the unsubstituted common isotopologues using $X$:

$$X_i = \frac{[\text{doubly substituted isotopologue}]}{[\text{unsubstituted isotopologue}]}. \tag{3.4}$$

The subscript $i$ refers to the clumped species. As an example, $X_{^{13}CH_3D} = \frac{[^{13}CH_3D]}{[CH_4]}$. We note that in typical studies of clumped isotopes, $R$ is used in place of $X$ — however, to avoid confusion over the use of $R$ for bulk isotopic compositions (as given in Equation (3.1)), we use a different symbol.

This excess or deficit of a clumped isotopologue vs. a random distribution is typically reported using capital delta notation:

$$\Delta_i = 1000 \times \left( \frac{X_i^{\text{sample}}}{X_i^{\text{random}}} - 1 \right) \tag{3.5}$$

In this equation, the "sample" refers to the actual measured relative concentration of the isotopologue, whereas the "random" refers to the composition that would be calculated assuming all isotopes are randomly distributed amongst all isotopologues. Both the $\alpha$ values (Eq. 3.3) and the $\Delta$ values (Eq. 3.5) can be related to expressions involving only the equilibrium constants for *certain* isotope exchange reactions. We describe these reactions in detail in the following section.

**Reactions describing the equilibrium isotope effects**

Equilibrium isotope effects as a function of temperature for bulk, site-specific, clumped, and clumped site-specific within and between various molecules are calculated from the equilibrium constant for an isotope-exchange reaction that describes the isotope effect of interest.

Equation (3.6) provides our generic notation for isotope exchange reactions between different phases or species $M$ vs $N$ (e.g., water vapor vs. liquid, or water vapor vs. methane gas). This reaction is used to describe bulk fractionation differences between isotopologues with one rare isotopic substitution (labelled with the *). The example of deuterium fractionation between methane and water is shown in Eq. (3.7).

$$^{*}M + N \rightleftharpoons M + {}^{*}N \tag{3.6}$$

$$CH_3D + H_2O \rightleftharpoons CH_4 + HDO \tag{3.7}$$

We demonstrate the validity of using a single exchange reaction to approximate bulk fractionations in Appendix A.3. For bulk fractionations involving propane two reactions of type (3.6) need to be considered, one for each of the two distinct sites. We represent the two sites by $C$ (center ethylene) and $T$ (terminal methyl) and write the following two reactions:

$$^{*}TCT + M \rightleftharpoons TCT + {}^{*}M \tag{3.8}$$

$$T^{*}CT + M \rightleftharpoons TCT + {}^{*}M. \tag{3.9}$$

Reaction (3.10) describes the site-specific isotope exchange of a heavy isotope (deuterium or carbon-13) labelled by * between the center ($C$) and terminal ($T$) sites of propane. An example for carbon-13 site-specific exchange in propane is shown in Eq. (3.11):

$$^{*}TCT \rightleftharpoons T^{*}CT \tag{3.10}$$

$$^{13}CH_3CH_2CH_3 \rightleftharpoons CH_3{}^{13}CH_2CH_3. \tag{3.11}$$

Here we assume that the three hydrogen atoms of the methyl group are indistinguishable. However, this is not always the case where there are *trans-* and *gauche-* deuterations (as in propane), which could be measured individually (e.g., by NMR). We discuss this in section 6.5. Note that any two of the three Equations (3.8), (3.9) & (3.10) together define both the bulk and the site-specific effect in propane, because the number of equations needed to determine the distribution of heavy isotopes among singly substituted species in two molecules is one less than the total number of sites in both molecules through closure.

To address clumped isotope effects, we divide our molecules of interest into three classes. The first class (dihydrogen, water, methane) has only one possible site for clumping (site $T$) and, as such, the clumped isotope effect can be uniquely defined for each possible combination of heavy isotopes (D + D, $^{13}$C + D, $^{17}$O + D and $^{18}$O + D). In such cases we only need to write one clumping reaction (3.12), where double star means two heavy isotopes at that site. The example for $^{13}$C+D clumping in methane is given in Eq. (3.13):

$$^*T + {}^*T \rightleftharpoons {}^{**}T + T \tag{3.12}$$

$$CH_3D + {}^{13}CH_4 \rightleftharpoons {}^{13}CH_3D + CH_4. \tag{3.13}$$

The second class has multiple sites for clumping that are all equivalent. The only molecule studied here of this class is ethane; however, other examples include cyclopropane (3 sites) and benzene (6 sites). For ethane we can write two clumped reaction types: on the same site (Eq. 3.14) and on the different sites (Eq. 3.15), where ethane is denoted as $TT$ to mark the two equivalent sites.

$$^*TT + {}^*TT \rightleftharpoons {}^{**}TT + TT \tag{3.14}$$

$$^*TT + {}^*TT \rightleftharpoons {}^*T^*T + TT \tag{3.15}$$

$$2\,CH_2D\text{--}CH_3 \rightleftharpoons CH_2D\text{--}CH_2D + C_2H_6 \tag{3.16}$$

Note that Eqs. (3.12), (3.14) and (3.15) can represent multiple isotopic equilibria reactions. For example, in the case of methane in addition to the reaction (3.13) there is also a reaction for D + D clumping. Eqs. 3.14 and 3.15 yield the clumped isotope effect for both D + D and D + $^{13}$C. However, reaction 3.14 cannot be written for $^{13}$C + $^{13}$C clumping since each site has only one carbon. Therefore, there are five distinct reactions describing the clumped isotope effects with two rare isotopic substitutions in ethane. If the *trans-* and *gauche-* rotamers in doubly deuterated ethane are distinguishable, Eq. (3.13) turns into two equations (one for each of the rotamers).

Finally, the third category considered are species with non-equivalent sites. Here the only molecule we consider in this category is propane, which has both equivalent and non-equivalent sites. The following species can be considered for D + D and $D + {}^{13}C$ clumping: $^{**}TCT$, $T^{**}CT$, $^*TC^*T$ and $^*T^*CT$. The latter two also appear for the $^{13}C + {}^{13}C$ clumping. All told, for propane we can write 11 distinct clumped isotope equilibrium reactions for two rare isotopes (again ignoring rotamers) — these are given in Appendix A.4. When writing these reactions, we adhere to the following rule: we ensure that the heavy atoms are located on the same sites between the left-hand side and the right-hand side of the equilibrium equation as is done in the example (3.17). This is done to isolate the clumping effect from site-specific effects. We write the equation as follows:

$$2\,{}^{13}CH_3\text{-}CH_2\text{–}CH_3 \;\rightleftharpoons\; {}^{13}CH_3\text{–}CH_2\text{–}{}^{13}CH_3 + C_3H_8. \tag{3.17}$$

To make this more concrete we also give a counterexample (3.18), where the clumped $^{13}C + {}^{13}C$ isotope effect is conflated with the site-specific $^{13}C$ isotope effect:

$$2\,{}^{13}CH_3\text{-}CH_2\text{–}CH_3 \;\rightleftharpoons\; {}^{13}CH_3\text{–}{}^{13}CH_2\text{–}CH_3 + C_3H_8. \tag{3.18}$$

**Relating observed isotopologue abundances to calculated isotope equilibrium constants**

There are two commonly employed ways to theoretically calculate values of fractionation factors ($\alpha$) at isotopic equilibrium. The first way is to calculate equilibrium isotope effects that describe isotope exchange reactions between all isotopologues of the given molecule types. This is an exact approach and was taken by Richet et al.[90] However, it involves a large number of calculations as all isotopologues (e.g., 216 of them for propane) must be considered. More recent approaches have avoided this and instead assumed that all isotopes are at a random distribution in each of the two species (phases), such that calculations of $\alpha$ can be approximated by using equilibrium constants for the exchange only between isotopologues with no substitutions and one rare substitution. Note that for molecules with nonequivalent sites (propane in our case) it is the average of substitutions in all (two for propane) sites that is assumed to be given by random distribution of isotopes. We take this latter approach here and demonstrate it is sufficiently accurate for our purposes in appendix A.3. Specifically, we calculate $\alpha^{eq}$ by normalizing the equilibrium constant ($K$) for a given isotope-exchange reaction relative to the equilibrium constant calculated with random distribution of isotopes among all isotopologues, $K^{\text{random}}$.

$$\alpha_i^{eq} = \frac{K_i}{K_i^{\text{random}}} \tag{3.19}$$

$K^{\text{random}}$ is the equilibrium constant that would be found for the infinitely high temperature limit. This normalization removes any dependence on the symmetry of the molecules, so that a deviation of $\alpha^{eq}$ from unity is solely due to isotope fractionation. Values of $\alpha^{eq}$ for reactions (3.6) are not exactly equal to the true values of fractionation factors $\alpha$ defined in Eq. (3.3). However, the approximate equality is true to high accuracy in all cases considered here because, at natural isotopic abundances, non-random concentrations of multiply substituted species do not significantly affect the value of $\alpha$ (see appendix A.3).

For propane we calculate the bulk fractionation factors using the two reactions between each site and the other species (Eqs. 3.8 and 3.9. The fractionation factors $\alpha_i^{eq}$ for each site are then averaged using appropriate weighting to yield the (approximate) total bulk fractionation between propane and another species. Appendix A.2 details this averaging procedure for both carbon-13 and deuterium bulk fractionations.

We now turn to clumped isotopes. The $\Delta$ values (Eq. 3.20) also depend on the abundance of heavy isotopes (see appendix A.5). Similarly to Eq. (3.19) we define $\Delta_X^{eq}$, which is independent of the heavy isotope abundances in the sample and expresses only the thermodynamic preference for the clumped isotope effect that results in doubly substituted isotopologue $X$. The value of $\Delta^{eq}$ is related to the equilibrium constants $K_X$ of the heavy isotope clumping reactions (3.12, 3.14, 3.15).

$$\Delta_X^{eq} = 1000 \times \ln \frac{K_X}{K_X^{\text{random}}} \tag{3.20}$$

For the clumped isotopic systems in which sites are equivalent (i.e., all species considered except propane), the $\Delta_X^{eq}$ values can be related to the corresponding $\Delta_X$ values in the infinite dilution limit (see appendix A.5). Eq. (3.20) does not appropriately approximate the $\Delta$ values given in Eq. (3.5) for propane (or any other molecule with non-equivalent sites). In such cases the contribution from the site-specific effects for singly substituted species needs to be included[91] in order to relate $\Delta_X^{eq}$ and $\Delta$. We discuss this in detail in appendix A.5.

The normalized equilibrium constant in Eq. (3.20) is equivalent to the ratio of the appropriate reduced partition function ratios (RPFRs) from reactions (3.12) for fractionation and reaction (3.10) for the site-specific effect:

$$\alpha_{\text{frac}}^{eq} = \frac{\text{RPFR}(^*N/N)}{\text{RPFR}(^*M/M)} \tag{3.21}$$

$$\alpha_{\text{site}}^{eq} = \frac{\text{RPFR}(T^*CT/TCT)}{\text{RPFR}(^*TCT/TCT)}. \tag{3.22}$$

Similarly, the normalized equilibrium constant for clumping (Eq. 3.20) is given by (3.23). Although we write (14C) explicitly for clumping in molecules with one site (9A), an analogous expression is valid for ethane (reactions 10A-C) and propane (reaction 11A and others in appendix A3).

$$\frac{K_X}{K_X^{\text{random}}} = \frac{\text{RPFR}(^{**}E/^*E)}{\text{RPFR}(^*E/E)} \tag{3.23}$$

The RPFRs in Eqs. (3.21), (3.22) and (3.23) are defined as follows:[92]

$$\text{RPFR}(M/N) = \frac{Q_M}{Q_N} \frac{\sigma_M}{\sigma_N} \prod_{\text{atoms}} \left( \frac{m_N}{m_M} \right)^{\frac{3}{2}} \tag{3.24}$$

where $Q$ are the partition functions for isotopologues $N$ and $M$, the product of the ratio of masses raised to the power of 3/2 runs over all isotopes that are different between $M$ and $N$, and $\sigma$ are the corresponding rotational symmetry numbers. Eq. (3.24) inherently normalizes out both the mass terms and the symmetry numbers in $Q$, such that at infinite temperature (i.e., in the classical limit) the RPFR is unity.

Throughout this work we report and discuss the thermodynamic preferences for fractionation, clumping and site-specific placement of heavy isotopes. We therefore report $\Delta_X^{eq}$ for clumped heavy isotope effects and $1000 \times \ln(\alpha_{N-M}^{eq})$ for heavy isotope fractionation between two phases throughout. We discuss the connection with the experimentally measurable quantities in Appendices A.3 and A.5. For heavy isotope fractionation at near-natural abundances of carbon-13 and deuterium $\alpha^{eq}$ (Eq. 3.19 approximates $\alpha$ (Eq. 3.3) to a very high precision for the molecules considered here (see appendix A.3). This is not the case for the clumped isotope effects as we show in appendix A.5. Nonetheless, the $\Delta_X^{eq}$ are independent of the heavy isotope abundance of the sample and express the true thermodynamic preference for the clumped isotope effect leading to isotopologue $X$. Most modern theoretical studies calculate $\Delta_X^{eq}$ and not the $\Delta_X$ values for these reasons.[86,87,93,94] In the last citation authors use a slightly different mathematical expression, but it is still based on the $K_X/K_X^{\mathrm{random}}$ ratio, just like Eq. (3.20). Finally, for all molecules considered here except for propane the $\Delta_X^{eq}$ values are sufficiently similar to the corresponding $\Delta_X$ values (the latter are up to a few per cent smaller at natural abundances of heavy isotopes), that all the observations discussed here for $\Delta_X^{eq}$ apply to $\Delta_X$ values as well.

In addition to the individual $\Delta_X^{eq}$ for each clumped isotopologue, we also report the combined total D + D and $^{13}$C + D effects for ethane and propane as well as the combined $^{13}$C + $^{13}$C effect for propane. These are reported in section 6.5 and calculated by weighing the individual $\Delta_X^{eq}$ as if isotopes were distributed randomly,[95] as detailed in Appendix A.6. Finally, we also provide $\Delta_{18}^{eq}$ for methane, $\Delta_{32}^{eq}$ for ethane and $\Delta_{46}^{eq}$ for propane in section 6.5, that combine all isotopologues with the same mass number, as detailed in Appendix A.6.

## 3.3 Methods

### Reduced partition function ratio (RPFR) calculations

The RPFRs are traditionally calculated within the harmonic approximation, assuming that the total partition function can be written as the product of vibrational, rotational and translational components; then the vibrations are approximated as harmonic, rotations as rigid, and both rotations and translations are treated as if they are classical. With these approximations, the RPFR of an isotopologue pair (starred over unstarred) can be written as:[90,92,96,97]:

$$\mathrm{RPFR}_{\mathrm{harmonic}} = e^{-\frac{E_0^*-E_0}{k_BT}} \prod_{j=1}^{a} \frac{\omega_j^*}{\omega_j} \times \frac{1-e^{-\frac{hc\omega_j}{k_BT}}}{1-e^{-\frac{hc\omega_j^*}{k_BT}}}. \tag{3.25}$$

In Eq. (3.25), $\omega_j$ is the harmonic frequency in wavenumbers (cm$^{-1}$) of the j$^{\text{th}}$ normal mode, $a$ is the total number of vibrational modes ($a = 3N - 5$ for linear molecules, $a = 3N - 6$ for non-linear molecules). The leading term ($E_0$) is the combined zero-point energy of all modes.

**PIMC calculations**

The path integral Monte Carlo (PIMC) method allows us to calculate the RPFRs without relying on the validity of the harmonic approximation.[86,93,98,99] Path-integral methods like PIMC or Path integral molecular dynamics (PIMD) fully take into account the anharmonicity of vibrations and rovibrational couplings, whereas the harmonic calculations do not. Neglecting these effects can lead to substantial errors for some isotopic equilibrium reactions[86,98–100], but not for others.[86,98]

Path-integral-based methods utilize the imaginary-time path-integral formalism[2] to sample the quantum Boltzmann statistics of the molecule. The remainder of this section describes the method, details the specific algorithm used for these computations and lists key parameters. The quantum statistics are sampled[3] by mapping each of the N (quantum) atoms onto n (classical) copies of that atom, yielding NP classical particles that interact via the modified potential:

$$V_n = \frac{m_n \omega_n^2}{2} \sum_{j=0}^{n-1} (q_{j+1} - q_j)^2 + V_n^{ext}(\boldsymbol{q}) \tag{3.26}$$

i.e., the potential part of the Hamiltonian given in Eq. (1.6). Thus, each quantum atom of the original molecule creates a ring polymer of $n$ beads, that are connected via harmonic springs. In the limit of infinite number of beads $n$, the partition function of the real quantum system is equal to the partition function of the fictitious P-times larger classical system. In practice we always use approximations to evaluate the potential energy and sufficiently large but finite $n$ is chosen. Then Monte Carlo (MC) or molecular dynamics (MD) is used to sample the classical system — here we employed the MC importance sampling method.[47]

While it is possible to take the nuclear exchange into account with PIMC,[4,101] we do not include this for two reasons. First, these effects are not expected to be important above 100K based on the estimate of the free ring-polymer radius of gyration[4] while all calculations presented here are for temperatures above 270 K. Secondly, including these effects reduces statistical precision of the MC sampling due to the so-called "sign problem,"[101] meaning much longer calculation times would be required with no expected gain in accuracy.

We used the direct scaled-coordinate estimator[102] to calculate the RPFR for every single heavy substitution relative to the lighter isotopologue and for every double substitution relative to the corresponding singly-substituted isotopologue. For each pair, the heavier of the two isotopologues was sampled with PIMC in Cartesian coordinates with hundreds of millions of Monte Carlo steps. Each Monte Carlo step consists of (1) moving the entire ring-polymers by a small random displacement in each coordinate and (2) $n/j$ staging moves[31] (rounded up to the nearest integer). The average displacement and staging length $j$ were set such that $40 \pm 2\%$ of all proposed staging moves are accepted to optimize sampling efficiency. Prior to any data collection, each sampling trajectory was equilibrated for $10^5$ Monte Carlo steps. Thereafter, ring-polymer configurations were sampled every 10 Monte Carlo steps. Aside from neglecting nuclear exchange, PIMC provides exact RPFRs for a specified potential energy surface (PES) in the limit of infinite sampling and infinite number of beads $n$. We choose the number of beads $n$ to balance the accuracy that improves as $n$ increases vs. the statistical uncertainty of sampling that increases with $n$. Note that the term accuracy here is relative to the method used to approximate the potential energy and not vs. the unknown true value. We ensure that in each case the convergence with respect to the number of beads $n$ for calculated RPFRs is within the statistical sampling error (standard error of the mean over samples) and report the latter as a measure of uncertainty.

**Diagonal Born-Oppenheimer correction**

All potential energy calculations in this work are done within the Born-Oppenheimer (BO) approximation.[103,104] This approximation separates the full Schrödinger equation into electronic and nuclear parts, greatly reducing the computational complexity and is employed, to our knowledge, in all calculations of equilibrium isotope effects. However, it has been shown that corrections to the BO approximation can substantially change calculated fraction factors for isotopes of hydrogen. For example, fractionation of deuterium between $H_2(g)$ and $H_2O(g)$ is affected by about 20‰ at room temperature[105] by the diagonal Born-Oppenheimer (DBO) correction, which is the lowest order perturbative correction to the BO-approximation. The DBO correction takes into account the dependence of the electronic wave function on the nuclear coordinates when calculating the nuclear kinetic-energy contribution. Higher order corrections are related to the excited electronic states[106] and, to our knowledge, have not been considered for equilibrium isotope effects as their contribution is presumed negligible for the molecules and temperature ranges considered in this book.

# Chapter 4

# CLUMPED ISOTOPE EFFECTS IN METHANE

53

## 4.1 Introduction

Methane is a product and reactant in atmospheric, hydrothermal, and magmatic chemical reactions and in microbial metabolisms. It is also a major component of commercial hydrocarbon deposits. A common first step in the study of methane in the environment, regardless of the application, is to constrain its source. A long-standing approach for this is to use the stable isotopic composition of a methane sample either through comparison of methane $^{13}C/^{12}C$ vs. D/H ratios to each other (given as $\delta^{13}C$ and $\delta D$ values),[107,108] to the concentration of alkanes gases (e.g., methane, ethane propane, and butane),[109] or to the stable isotopic composition of larger alkane gases.[110] The measurement of molecules of methane with multiple rare, heavy isotopes such as $^{13}CH_3D$ and $CH_2D_2$ has provided a new way to study the formational conditions of methane.[111–113] The abundance of these so-called "clumped" isotopologues for an isotopically equilibrated system relative to that expected for a random distribution of isotopes among all methane molecules is a monotonic a function of temperature.[91] Thus, the measurement of methane clumped isotopic compositions (relative to a random isotopic distribution) can in principle be used as a geothermometer and to study departures of samples from isotopic equilibrium. Applications of methane clumped isotopes studies include the determination of apparent formation (or re-equilibration) temperatures of methane in subsurface reservoirs and to fingerprint abiotic, biogenic, and thermogenic methane.[113–123]

These capabilities arise from the ability to precisely measure (order per mil) the relative abundances of unsubstituted ($CH_4$), singly-substituted ($CH_3D$, $^{13}CH_4$) and multiply- substituted isotopologues of methane ($^{13}CH_3D$, $CH_2D_2$) using either high-resolution gas-source isotope-ratio mass spectrometers[111,113,124,125] or laser adsorption spectrometers.[112]

Regardless of the technique, measurements are performed relative to commercial high-purity methane "working gases," which have an *a priori* unknown clumped isotope composition. As a result, measured methane clumped isotopic compositions are not inherently anchored to an external reference frame such as one set by international standards (which are not available) or set by theoretical expectations to the temperature dependence of methane clumping. To report methane clumped-isotopic compositions that are anchored to a thermodynamic reference frame and, thus, be able to calculate clumped-isotope based formation or apparent temperatures and compare measurements among different l aboratories, p ast s tudies h ave combined experiment and theory to place measured clumped-isotopic compositions into a reference frame anchored by theoretical expectations of the temperature dependence of

$^{13}CH_3D$ or $CH_2D_2$ concentrations vs. their expected concentrations for a system in isotopic equilibrium with a random distribution of isotopes.[111–113,120] To accomplish this, previous studies isotopically equilibrated methane isotopologues at temperatures greater than 150°C in the presence of catalysts that promote C-H bond activation and hydrogen isotope exchange. The measured differences between samples equilibrated at different temperatures were then compared to statistical mechanical-based calculation of these expected differences.[111–113] All measurements of clumped methane compositions are based on this approach and are performed on a lab-by-lab basis. The accuracy of such "heated gas" calibrations and, thus, measured methane clumped isotopic compositions and apparent temperatures depends on the accuracy of the calculation, the experiments, and the isotopic measurements.

**Isotope exchange reactions and nomenclature**

Two clumped methane isotopologues ($^{13}CH_3D$ and $CH_2D_2$) have been measured at precisions necessary to calculate clumped-isotope based temperatures at useful precisions ($\pm < 25°C$) and temperatures $< 200°C$ for samples with natural abundances of stable isotopes. The abundances of these species for a given measurement are reported using $\Delta$ notation[91] (Eq. 3.5) such that:

$$\Delta_{^{13}CH_3D} = 1000 \times \left( \frac{[^{13}CH_3D]/[CH_4]}{[^{13}CH_3D]^*/[CH_4]^*} - 1 \right) \tag{4.1}$$

and

$$\Delta_{CH_2D_2} = 1000 \times \left( \frac{[CH_2D_2]/[CH_4])}{[CH_2D_2]^*/[CH_4]^*} - 1 \right) \tag{4.2}$$

In Eqs. (4.1-4.2), the square brackets denote concentrations relative to all other methane isotopologues and the * (star) denotes the calculated concentration of an isotopologue assuming all isotopes of carbon and hydrogen are randomly distributed among all isotopologues.[91] These $\Delta$ values can be related to the following equilibrium isotope-exchange reactions:

$$^{13}CH_4 + CH_3D \rightleftharpoons CH_4 + {}^{13}CH_3D \tag{4.3}$$

and

$$2\,CH_3D \rightleftharpoons CH_4 + CH_2D_2 \tag{4.4}$$

with $K_{^{13}CH_3D}$ and $K_{CH_2D_2}$ describing the equilibrium constants for Eqs. (4.3) and (4.4), respectively.

For isotopically equilibrated systems, $\Delta$ and $K$ values are related through the following equations:[91]

$$\Delta_{^{13}CH_3D} \cong \Delta^{eq}_{^{13}CH_3D} = 1000 \times \ln\left(K_{^{13}CH_3D}\right) \tag{4.5}$$

and

$$\Delta_{CH_2D_2} \cong \Delta^{eq}_{CH_2D_2} = 1000 \times \ln\left(\frac{8}{3}K_{CH_2D_2}\right). \tag{4.6}$$

The 8/3 value is present in Eq. (4.6) due to the differing symmetry numbers of the various methane isotopologues in Eq. (4.4). The approximate signs are present because we assume that the concentrations of the $^{13}CH_4$ and $CH_3D$ isotopologues are equal to values expected for a stochastic isotopic distribution. This is only approximately true, but this approximation is valid for our purposes[111] given both the measurement precisions ($\pm 1$ s.e.) that will be reported below for $\Delta_{^{13}CH_3D}$ ($\pm 0.25$-$0.3$‰) and $\Delta_{CH_2D_2}$ ($\pm 1$-$1.5$‰) as well as typical $\delta^{13}C$ and $\delta D$ ranges of environmental samples ($\sim 70$‰ and $\sim 500$‰, respectively).

We note that an additional parameter that was used for the first methane clumped isotope measurements is $\Delta_{18}$.[111] This represents the combined measurements of $^{13}CH_3D$ and $CH_2D_2$ vs. $CH_4$ compared to a random isotopic distribution.[111] $\Delta 18$ values are largely equivalent to $\Delta_{^{13}CH_3D}$ values because 98% of the cardinal mass-18 methane isotopologues are $^{13}CH_3D$ and only 2% are $CH_2D_2$.

The key point for our purposes here is that the measured $\Delta$ quantities are directly related to temperature-dependent equilibrium isotope exchange reactions for isotopically equilibrated systems (i.e., in homogeneous phase equilibrium). Thus, if samples can be isotopically equilibrated at known temperatures and the theoretically expected differences calculated, then the $\Delta$ value of samples can be converted into apparent temperatures based on well-understood quantum-statistical-mechanical theories regardless of the clumped-isotopic composition of the reference gas used during measurements.

**Previous experimental and theoretical determinations of the temperature dependence of $\Delta$ values for isotopically equilibrated systems**

Experimental calibrations and temperature dependencies of $\Delta_{^{13}CH_3D}$ and $\Delta_{CH_2D_2}$ for isotopically equilibrated systems have been conducted at temperatures above $150°C$[119] and $300°C$,[113] respectively, and above $200°C$ for $\Delta_{18}$ values.[111] In contrast, formation temperatures of biogenic gases on earth are typically below $80°C$[126,127] while thermogenic gases are thought to begin forming as low as $60°C$.[128] Thus, the potential range of expected gas-formation temperatures in nature is commonly outside of these calibrated ranges. This requires extrapolation of calibrations to lower temperatures and higher $\Delta$ values to calculate clumped-isotope based temperatures. Stolper and co-workers[111] calibrated equilibrium $\Delta_{18}$ values at four temperatures (200, 300, 400, and 500°C) using a nickel-based catalyst that represented a total measured range in $\Delta_{18}$ of 1.8‰ (quoted internal precision of $\pm 0.25$-$0.3$‰, $1\sigma$ and external precision of $\pm 0.25$-$0.3$‰, $1\sigma$). Following this, Ono and co-workers[112] calibrated equilibrium $\Delta_{^{13}CH_3D}$ values at three temperature values over the range of 200 to 400°C using a platinum-based catalyst that represented a total measured range in $\Delta_{^{13}CH_3D}$ of about 1.4‰ (quoted $1\sigma$ internal precision $\sim 0.1$-$0.15$‰, and analytical $1\sigma$ precision of $\pm 0.35$‰). In the same laboratory, Wang and co-workers[119] performed a similar calibration using platinum catalysts at three temperature values over 150 to 250°C and a total measured range in $\Delta_{^{13}CH_3D}$ of about 1.2‰. They additionally measured a sample at 400°C, but this data point was not included in their calibration because it did not fit with the expected theoretical temperature dependence. It was proposed that the sample may have been compromised by potential quench effects. Finally, Young and co-workers[113] calibrated equilibrium $\Delta_{^{13}CH_3D}$ and $\Delta_{CH_2D_2}$ values at three temperatures (300, 400, 500°C) using a platinum-based catalyst representing a total measured range in $\Delta_{^{13}CH_3D}$ of 1.0‰ and $\Delta_{CH_2D_2}$ of about 2.2‰ (quoted internal $1\sigma$ precision ±0.15‰ and ±0.35‰, respectively, and ±0.3‰ and ±1.0‰ external precision,[116] respectively). Wang et al.[129] also used $\gamma$-$Al_2O_3$ catalysts to equilibrate $\Delta_{^{13}CH_3D}$ values of methane at 25 and 100°C. Their total measured $\Delta_{^{13}CH_3D}$ range is 2.23‰ with analytical precisions of generally ±0.5‰ ($1\sigma$).

The lack of samples equilibrated at temperatures <150°C, despite expectations that biogenic and thermogenic gases could form at such temperatures, is due to the usage of catalysts (nickel- and platinum-based) that are sluggish at temperatures <150°C over laboratory timescales. For example, the calibration of the equilibrium value for $\Delta_{^{13}CH_3D}$ at 150°C (representing the lowest clumped methane calibration temperature reported in the above studies) is based on a single experiment that was allowed to equilibrate for 110 days.[119] The ability to extend calibrations to lower temperatures using methane equilibrated in the laboratory would allow for more detailed comparisons between theory and experiment and allow apparent clumped-isotope based temperatures to be calculated based on interpolation of calibrations as opposed to extrapolations.

Previous calculation of equilibrium $\Delta_{^{13}CH_3D}$ and $\Delta_{CH_2D_2}$ values are based on one of two theoretical approaches: (i) The Bigeleisen and Mayer/Urey model[92,96] (BMU),[87,95,113,119,130,131] which in practice involves calculations of so-called reduced partition function ratios (RPFRs) using a harmonic approximation for the treatment of the vibrational partition function and classical expressions for rotational and translational partition functions; and (ii) Path Integral Monte Carlo (PIMC) simulations that avoid the major approximations in the BMU model yielding a fully anharmonic and quantum mechanical description of the partition function ratios.[86]

Both approaches require independent computations of the electronic PES for methane, which are typically taken from electronic structure calculations based on density functional theory (DFT) or more accurate ab initio wavefunction theories, such as coupled-cluster theory. Differences in previous calculation of equilibrium $\Delta_{^{13}CH_3D}$ values as a function of temperature based on the BMU model using harmonic frequencies are comparable to the typical internal precision of $\Delta_{^{13}CH_3D}$ measurements ($\leq 0.2$‰ for T $\geq 0$°C)[87,130,131]. Cao and Liu[95] initially assessed the effect of the anharmonic corrections to $\Delta_{^{13}CH_3D}$ and found deviations (up to ~0.2‰) that are comparable to typical internal precision of $\Delta^{13}CH_3D$ measurements. A later study[130] applied a series of corrections to harmonic RPFRs to account for the effects of anharmonicity and many of the other major approximations inherent to the BMU model using computed isotopologue-specific molecular constants[90,132] and found overall smaller differences in computed $\Delta_{^{13}CH_3D}$ values relative to uncorrected values based on harmonic RPFRs, but the differences may be systematic in nature (i.e., +0.04‰ at 0°C to +0.08‰ at 725°C).

Webb and Miller[86] performed calculation of $\Delta_{^{13}CH_3D}$ using both PIMC and BMU approaches (with and without anharmonic corrections) based on the same computed electronic PES[133] from 27 to 327°C. In that work, the PIMC calculations are the most rigorous and accurate approach to calculating $\Delta$ values for isotopically equilibrated systems. Calculations of $\Delta_{^{13}CH_3D}$ using both PIMC and BMU-harmonic approaches yielded similar results over the temperature ranges studied (i.e., all are within $\leq 0.06‰$ over 27-327°C). However, they illustrated that the apparent agreement between the BMU-harmonic and PIMC calculations arises due to a precise cancelation of errors in the harmonically computed RPFRs during computation of the equilibrium constant. They further demonstrated that an anharmonic correction to the vibrational zero point energy resulted in comparatively worse agreement (e.g., 0.2 to 0.4‰ differences in $\Delta_{^{13}CH_3D}$ relative to PIMC).[86] It is important to note that precise error cancelation between PFRs was not universally observed by Webb and Miller. For example, in the isotope exchange reaction describing position-specific isotope abundances for isotopically equilibrated system between $^{14}N^{15}NO$ and $^{15}N^{14}NO$, an anharmonic correction did yield overall better agreement with PIMC results. This indicates that the partial corrections to BMU approaches should be judiciously applied in the absence of converged PIMC calculations. calculation of equilibrium $\Delta_{CH_2D_2}$ values as a function of temperature have been performed in two studies based solely on the harmonic BMU model.[87,113] The calculated $\Delta_{CH_2D_2}$ values from these two studies as a function of temperature are similar (<0.45‰ for temperatures $\geq 0°C$).

Here, we provide an experimental calibration of equilibrium $\Delta_{^{13}CH_3D}$ and $\Delta_{CH_2D_2}$ values from 1-500°C and compare it to new theoretical computations of equilibrium $\Delta_{^{13}CH_3D}$ and $\Delta_{CH_2D_2}$ as a function of temperature using PIMC methods[86,93] and BMU calculations based on the same electronic PES to facilitate direct comparison. To achieve isotopic equilibrium on laboratory time scales, we use a $\gamma$-$Al_2O_3$ catalyst to equilibrate methane from 1 to 165°C and a nickel-based catalyst for higher temperatures (250 to 500°C). We then compare these results to the expected differences using different theoretical approaches for computing clumped methane compositions (i.e., PIMC and BMU). We show that the theoretical and experimental measurements are in 1:1 agreement from 1 to 500°C and thus provide a calibration for both $\Delta_{^{13}CH_3D}$ and $\Delta_{CH_2D_2}$ over this temperature range validated by both experiment and theory.

| | 300 K | | | 400 K | | | 500 K | | | 600 K | |
|---|---|---|---|---|---|---|---|---|---|---|---|
| $n$ | $\frac{CH_3D}{CH_4}$ | $\frac{CH_2D_2}{CH_3D}$ | $n$ | $\frac{CH_3D}{CH_4}$ | $\frac{CH_2D_2}{CH_3D}$ | $n$ | $\frac{CH_3D}{CH_4}$ | $\frac{CH_2D_2}{CH_3D}$ | $n$ | $\frac{CH_3D}{CH_4}$ | $\frac{CH_2D_2}{CH_3D}$ |
| 32 | 113.380 | 43.346 | 24 | 54.860 | 20.766 | 20 | 35.919 | 13.535 | 18 | 27.231 | 10.244 |
| 64 | 117.250 | 44.832 | 48 | 56.238 | 21.296 | 40 | 36.581 | 13.790 | 35 | 27.714 | 10.424 |
| 128 | 118.300 | 45.216 | 96 | 56.614 | 21.430 | 80 | 36.759 | 13.857 | 70 | 27.810 | 10.461 |
| 256 | 118.520 | 45.315 | 192 | 56.701 | 21.461 | 160 | 36.797 | 13.872 | 140 | 27.841 | 10.471 |
| 384 | 118.610 | 45.332 | 288 | 56.719 | 21.477 | 240 | 36.816 | 13.873 | 210 | 27.849 | 10.473 |
| 512 | 118.590 | 45.322 | 384 | 56.727 | 21.475 | 320 | 36.817 | 13.876 | 280 | 27.846 | 10.473 |

Table 4.1: Convergence of the calculated PFRs for isotopologue pairs $CH_3D/CH_4$ and $CH_2D_2/CH_3D$ with the number of beads $n$.

## 4.2 Methods

The methods used in the experimental portion of this study are detailed in,[94] including the origin, purification and equilibration of methane, the precise isotopic measurements via mass spectrometry with ThermoFisher 253 Ultra instrument, and the conversion of measured peaks to $\delta D$, $\delta^{13}C$, $\Delta_{13CH_3D}$ and $\Delta_{CH_2D_2}$ values. Here we only describe how the reduced partition function ratios (RPFRs) are computed using the PIMC.[86,93] The electronic PES of methane is taken from Lee and co-workers[133] and is computed at the CCSD(T) level of theory.

The Path Integral Monte Carlo (PIMC) technique is described in section 3.3. A direct scaled-coordinate estimator[102] is used to calculate the RPFRs. Heavy isotopologue configurations are sampled with PIMC in Cartesian coordinates with an explicit staging transformation.[31] The staging length, $j$, is set such that 38-42% of all proposed staging moves are accepted. Prior to any data collection, each sampling trajectory is equilibrated for $10^5$ MC steps, with $n/j$ staging moves (rounded up to the nearest integer) attempted per MC step. Thereafter, ring-polymer configurations are sampled every 10 MC steps. The total number of MC moves for each RPFR calculation is $2 \times 10^8$. There are two primary sources of error in the PIMC calculations, aside from any errors due to the PES. The first is systematic error related to convergence of the RPFRs with the number of beads; this error vanishes in the limit of infinite beads (Eq. 1.5). The second is statistical error related to sampling of the direct scaled-coordinate estimator for the PFRs; this error vanishes in the limit of infinite sampling. We determine the number of beads employed in the PIMC calculations based on explicit convergence tests for the individual PFRs (see Fig. 4.1) and summarized in Table 4.1. All errors reported for the PIMC calculations reflect standard errors related to statistical uncertainty with the Monte Carlo sampling method.

## 4.3 Results

Results of the PIMC calculations are presented in Table 4.2. Values of $\Delta^{eq}_{13CH_3D}$ and $\Delta^{eq}_{CH_2D_2}$ have been computed over a temperature range of -3.15 to 527.85°C (270 to 800K). The errors ($\pm 1$ s.e.) on individual PIMC computations are $\leq 0.03$‰ for $\Delta^{eq}_{13CH_3D}$ and $\leq 0.33$‰ for $\Delta^{eq}_{CH_2D_2}$.

| T (°C) | n | RPFR | | | | $^{13}CH_3D$ | | $CH_2D_2$ | |
|---|---|---|---|---|---|---|---|---|---|
| | | $CH_3D$ | $^{13}CH_4$ | $^{13}CH_3D$ | $CH_2D_2$ | $\Delta^{eq}$ | 1 s.e. | $\Delta^{eq}$ | 1 s.e. |
| 1.2 | 414 | 156.6 | 1.271 | 1.280 | 60.11 | 6.598 | 0.031 | 23.28 | 0.13 |
| 6.9 | 408 | 146.6 | 1.267 | 1.275 | 56.22 | 6.477 | 0.021 | 22.18 | 0.22 |
| 16.9 | 396 | 131.4 | 1.261 | 1.269 | 50.29 | 6.036 | 0.030 | 20.44 | 0.33 |
| 26.9 | 381 | 118.6 | 1.255 | 1.262 | 45.33 | 5.764 | 0.029 | 18.99 | 0.26 |
| 36.9 | 372 | 107.7 | 1.250 | 1.256 | 41.13 | 5.441 | 0.026 | 17.96 | 0.24 |
| 46.9 | 360 | 98.50 | 1.245 | 1.251 | 37.55 | 5.201 | 0.023 | 16.52 | 0.24 |
| 50.5 | 357 | 95.47 | 1.243 | 1.249 | 36.38 | 5.081 | 0.029 | 16.10 | 0.10 |
| 56.9 | 351 | 90.57 | 1.240 | 1.246 | 34.48 | 4.934 | 0.023 | 15.18 | 0.21 |
| 66.9 | 342 | 83.65 | 1.235 | 1.241 | 31.83 | 4.671 | 0.020 | 14.70 | 0.22 |
| 75.7 | 333 | 78.33 | 1.232 | 1.237 | 29.77 | 4.467 | 0.018 | 13.34 | 0.09 |
| 76.9 | 333 | 77.68 | 1.231 | 1.237 | 29.52 | 4.484 | 0.024 | 13.21 | 0.17 |
| 86.9 | 324 | 72.39 | 1.227 | 1.233 | 27.49 | 4.246 | 0.019 | 12.59 | 0.20 |
| 96.9 | 315 | 67.78 | 1.224 | 1.229 | 25.71 | 4.054 | 0.018 | 11.32 | 0.21 |
| 126.9 | 294 | 56.72 | 1.214 | 1.218 | 21.48 | 3.504 | 0.017 | 9.68 | 0.16 |
| 127.8 | 294 | 56.43 | 1.214 | 1.218 | 21.36 | 3.483 | 0.016 | 9.34 | 0.13 |
| 151.9 | 279 | 49.88 | 1.207 | 1.211 | 18.86 | 3.123 | 0.013 | 8.22 | 0.13 |
| 165.4 | 273 | 46.84 | 1.203 | 1.207 | 17.70 | 2.947 | 0.017 | 7.45 | 0.07 |
| 176.9 | 267 | 44.54 | 1.201 | 1.204 | 16.82 | 2.809 | 0.016 | 7.02 | 0.10 |
| 201.9 | 255 | 40.28 | 1.195 | 1.198 | 15.19 | 2.522 | 0.014 | 5.69 | 0.11 |
| 226.9 | 243 | 36.81 | 1.190 | 1.193 | 13.88 | 2.303 | 0.016 | 5.14 | 0.12 |
| 251.9 | 234 | 33.96 | 1.186 | 1.188 | 12.79 | 2.042 | 0.011 | 4.44 | 0.11 |
| 276.9 | 225 | 31.57 | 1.182 | 1.184 | 11.89 | 1.867 | 0.009 | 3.90 | 0.13 |
| 301.9 | 216 | 29.56 | 1.179 | 1.181 | 11.12 | 1.684 | 0.009 | 3.27 | 0.13 |
| 326.9 | 207 | 27.84 | 1.175 | 1.177 | 10.47 | 1.536 | 0.012 | 2.98 | 0.11 |
| 351.9 | 201 | 26.37 | 1.173 | 1.174 | 9.914 | 1.399 | 0.007 | 2.65 | 0.12 |
| 376.9 | 195 | 25.09 | 1.170 | 1.171 | 9.430 | 1.288 | 0.010 | 2.33 | 0.10 |
| 401.9 | 189 | 23.97 | 1.168 | 1.169 | 9.006 | 1.169 | 0.008 | 1.93 | 0.10 |
| 426.9 | 183 | 22.99 | 1.165 | 1.167 | 8.635 | 1.066 | 0.007 | 1.75 | 0.10 |
| 451.9 | 177 | 22.11 | 1.163 | 1.164 | 8.306 | 0.966 | 0.006 | 1.60 | 0.08 |
| 476.9 | 174 | 21.34 | 1.161 | 1.163 | 8.013 | 0.892 | 0.007 | 1.46 | 0.09 |
| 501.9 | 168 | 20.65 | 1.160 | 1.161 | 7.752 | 0.828 | 0.006 | 1.15 | 0.06 |
| 526.9 | 165 | 20.02 | 1.158 | 1.159 | 7.519 | 0.752 | 0.007 | 1.29 | 0.09 |

Table 4.2: Results of PIMC calculations of the reduced partition function ratios (RPFRs) and $\Delta^{eq}$ values.

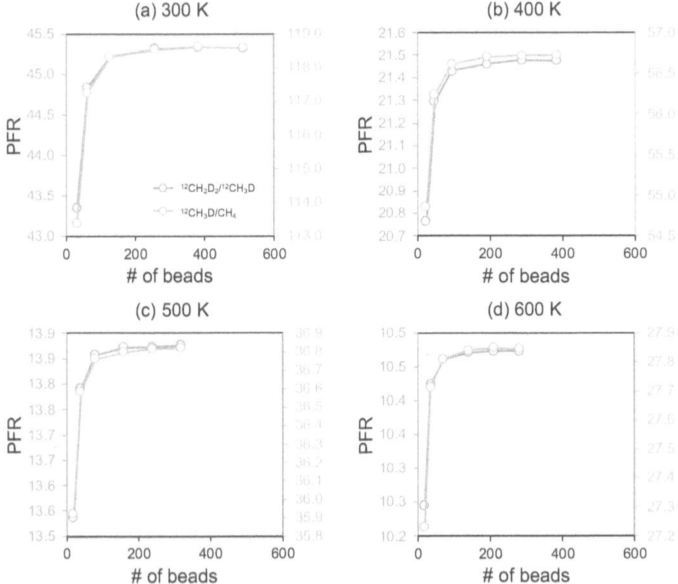


**Figure SI.5(a-d):** PIMC calculations: Convergence of $^{12}CH_3D/CH_4$ (blue) and $^{12}CH_2D_2/CH_3D$ (grey) PFR's with the number of beads. The number of beads for the other temperatures is determined by interpolation between the

**Figure 4.1: Convergence of PFR's with the number of beads** for $CH_3D/CH_4$ (blue) and $CH_2D_2/CH_3D$ (grey). The number of beads for the calculations at intermediate temperatures is determined by interpolating between the second last points on each panel. Data are also shown in Table 4.1.

Polynomial fits to $\Delta$ values as a function of $T^{-1}$ ($6^{th}$ and $7^{th}$ order with a forced intercept through 0‰ at infinite temperature) have been applied to the PIMC results to allow interpolation between computed temperatures and extrapolation above the highest computed temperature:

$$\Delta_{^{13}CH_3D} \cong 1000 \times \ln\left(K_{^{13}CH_3D}\right) = \frac{1.47348 \times 10^{19}}{T^7} - \frac{2.08648 \times 10^{17}}{T^6} + \frac{1.19810 \times 10^{15}}{T^5}$$
$$- \frac{3.54757 \times 10^{12}}{T^4} + \frac{5.54476 \times 10^9}{T^3} - \frac{3.49294 \times 10^6}{T^2} - \frac{8.89370 \times 10^2}{T}$$

$$(4.7)$$

$$\Delta_{CH_2D_2} \cong 1000 \times \ln\left(\frac{8}{3}K_{CH_2D_2}\right) = -\frac{9.67634 \times 10^{15}}{T^6} + \frac{1.71917 \times 10^{14}}{T^5} - \frac{1.24819 \times 10^{12}}{T^4}$$
$$+ \frac{4.30283 \times 10^9}{T^3} - \frac{4.48660 \times 10^6}{T^2} + \frac{1.86258 \times 10^3}{T}.$$

$$(4.8)$$

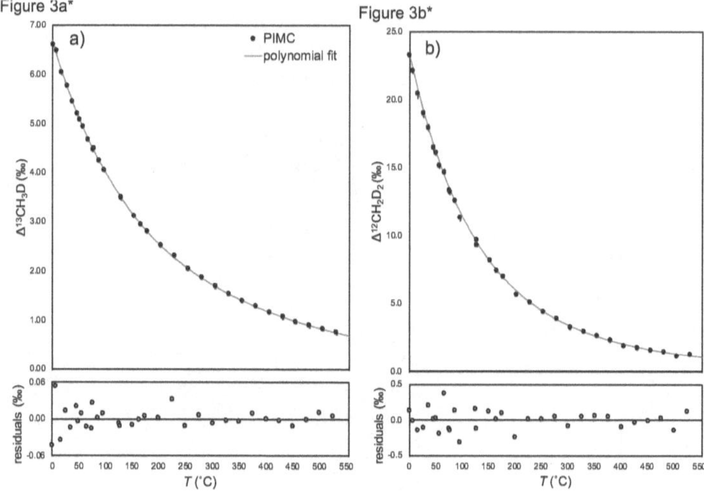

Figure 4.2: **Polynomial fit to the PIMC calculations** of $\Delta^{eq}_{^{13}CH_3D}$ (panel a) and $\Delta^{eq}_{CH_2D_2}$ (panel b). The residuals from the fit are randomly distributed. Error bars on PIMC calculations (1 s.e.) are smaller than the data points. Best-fit polynomials are given in Eqs. (4.7) and (4.8).

Values computed from these equations are strictly valid over 270 to 800K but are also likely valid above 800 K due to the requirement that these values (as defined) must approach 0‰ at the high-temperature-limit. Note that the range in values of $\Delta_{^{13}CH_3D}$ and $\Delta_{CH_2D_2}$ extrapolated above 800K to the high temperature limit are $\leq 0.75$‰ and $\leq 1.3$‰, respectively. Computed values from the polynomial fits are shown in Fig 4.2 along with residuals. The residuals of the polynomial fit are randomly distributed and the $\pm 1\sigma$ of the residuals ($\pm 0.02$‰ and $\pm 0.14$‰ for the $\Delta^{eq}_{^{13}CH_3D}$ and $\Delta^{eq}_{CH_2D_2}$ polynomial fits, respectively) are comparable to the errors in the calculations at any given temperature (see Table 4.2).

Table 4.3 contains the differences between the PIMC and the BMU approaches in $\delta$-notation: $\delta \mathrm{RPFR}_{BMU\text{-}PIMC} = 1000 \times (\mathrm{RPFR}_{BMU}/\mathrm{RPFR}_{PIMC} - 1)$, and $\delta K_{BMU\text{-}PIMC} = 1000 \times K_{BMU}/K_{PIMC} - 1)$ (reported in units of ‰). Both the equilibrium constants ($K_{^{13}CH_3D}$ and $K_{CH_2D_2}$) and the equilibrium clumped compositions $\Delta^{eq}_{^{13}CH_3D}$ and $\Delta^{eq}_{CH_2D_2}$ (derived from $K$'s using Eqs. 4.1 and 4.2) are comparable between the PIMC and BMU approaches (Table 4.3). For example, values of $\Delta^{eq}_{^{13}CH_3D}$ computed using the BMU model are within $\leq 0.10$‰ of the PIMC values over all computed temperatures (-3-527°C). Similarly, values of $\Delta^{eq}_{CH_2D_2}$ computed using the BMU model are within $\leq 0.37$‰ of the PIMC values. Despite this agreement in equilibrium constants, the BMU-based RPFRs are systematically higher than the

| T (°C) | $\delta$RPFR(BMU-PIMC) (‰) | | | | $\delta K$ (‰) | |
|---|---|---|---|---|---|---|
| | $CH_3D$ | $^{13}CH_4$ | $^{13}CH_3D$ | $CH_2D_2$ | $^{13}CH_3D$ | $CH_2D_2$ |
| -3.1 | 94.1 | 5.5 | 100 | 198 | -0.08 | 0.38 |
| 1.2 | 93.0 | 5.4 | 99 | 195 | -0.01 | 0.14 |
| 6.9 | 91.1 | 5.3 | 97 | 191 | -0.10 | 0.23 |
| 16.9 | 87.6 | 5.1 | 93 | 183 | 0.00 | 0.30 |
| 26.9 | 84.5 | 4.9 | 90 | 176 | -0.06 | 0.21 |
| 36.9 | 82.0 | 4.8 | 87 | 171 | -0.03 | -0.16 |
| 46.9 | 79.3 | 4.6 | 84 | 165 | -0.07 | -0.01 |
| 50.5 | 78.3 | 4.6 | 83 | 163 | -0.05 | -0.03 |
| 56.9 | 76.5 | 4.5 | 81 | 159 | -0.06 | 0.16 |
| 66.9 | 74.6 | 4.4 | 79 | 154 | -0.04 | -0.44 |
| 75.7 | 72.3 | 4.2 | 77 | 150 | -0.04 | 0.03 |
| 76.9 | 72.0 | 4.2 | 76 | 149 | -0.08 | 0.05 |
| 86.9 | 70.3 | 4.1 | 75 | 145 | -0.06 | -0.24 |
| 96.9 | 68.0 | 4.0 | 72 | 141 | -0.06 | 0.18 |
| 126.9 | 62.7 | 3.7 | 67 | 129 | -0.04 | -0.33 |
| 127.8 | 62.4 | 3.7 | 66 | 129 | -0.04 | -0.05 |
| 151.9 | 58.8 | 3.4 | 62 | 121 | -0.03 | -0.31 |
| 165.4 | 56.8 | 3.3 | 60 | 117 | -0.04 | -0.21 |
| 176.9 | 55.3 | 3.3 | 59 | 113 | -0.04 | -0.30 |
| 201.9 | 51.9 | 3.1 | 55 | 107 | -0.04 | 0.05 |
| 226.9 | 49.3 | 2.9 | 52 | 101 | -0.07 | -0.21 |
| 251.9 | 46.8 | 2.7 | 50 | 95 | -0.03 | -0.20 |
| 276.9 | 44.4 | 2.6 | 47 | 91 | -0.04 | -0.23 |
| 301.9 | 42.2 | 2.5 | 45 | 86 | -0.03 | -0.08 |
| 326.9 | 40.4 | 2.4 | 43 | 82 | -0.04 | -0.20 |
| 351.9 | 38.5 | 2.3 | 41 | 78 | -0.03 | -0.22 |
| 376.9 | 36.9 | 2.2 | 39 | 75 | -0.05 | -0.19 |
| 401.9 | 35.3 | 2.1 | 37 | 72 | -0.04 | -0.04 |
| 426.9 | 33.8 | 2.0 | 36 | 69 | -0.03 | -0.09 |
| 451.9 | 32.6 | 1.9 | 35 | 66 | -0.02 | -0.12 |
| 476.9 | 31.4 | 1.9 | 33 | 64 | -0.02 | -0.14 |
| 501.9 | 30.1 | 1.8 | 32 | 61 | -0.03 | 0.03 |
| 526.9 | 29.2 | 1.7 | 31 | 59 | -0.02 | -0.24 |

Table 4.3: Comparison between approximate BMU calculations and the PIMC methods in $\delta$ notation.

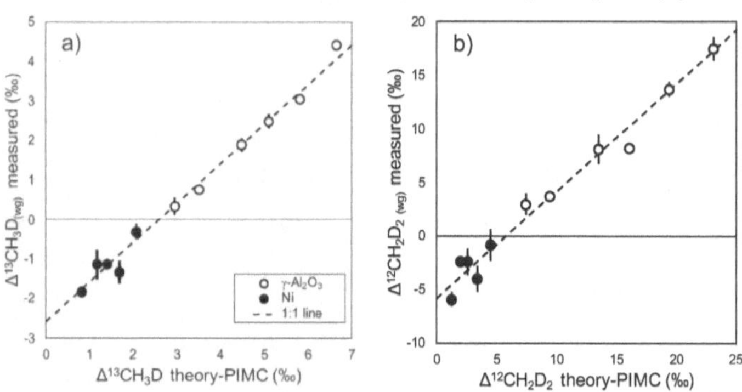

Figure 4.3: **Comparison of PIMC calculations to experiment** $1000 \times \ln\left(\Delta_{(wg)}/1000 + 1\right)$ values from experiments (y-axis) vs. $1000 \times \ln(K_{^{13}CH_3D})$ (panel a) and $1000 \times \ln\left(8/3 \times K_{CH_2D_2}\right)$ (panel b) values from the PIMC calculations obtained at the experimental temperatures. White circles indicate $\gamma$-$Al_2O_3$ and black circles indicate Ni experiments. Error bars for replicated experimental data points from this study represent either the $\pm 1$ s.e. of replicates or the expected $\pm 1$ s.e. based on the observed external precision of standards and the number of experimental replicates (i.e., $\sigma_{external}/\sqrt{n}$, where n is the number of experimental replicates), whichever is larger. Error bar on one experimental data point that has not been replicated (the $\Delta_{CH_2D_2}$ value at 350 °C) represents $\pm 1$ s.e. internal precision. Error bars in the PIMC calculations (x axis error bars, $\pm$ 1 s.e.) are smaller than the symbols.

PIMC RPFRs. The $\delta$ RPFR$_{BMU\text{-}PIMC}$ are as high as 5-6‰ for $^{13}CH_4$ & $^{13}CH_3D$ and 29-95‰ for the CH$_3$D & CH$_2$D$_2$ over the computed temperature range.

## 4.4   Discussion

### Comparison to the experiment

The $\Delta_{^{13}CH_3D}$ and $\Delta_{CH_2D_2}$ values for each equilibrated sample can only be obtained in reference to an arbitrary chosen reference frame that is not rooted in thermodynamics or internationally recognized standards (which do not exist for methane clumped measurements). Therefore, the measurements are performed in reference to the working gas (house methane), whose composition (i.e., $\Delta_{^{13}CH_3D}$ and $\Delta_{CH_2D_2}$ values) is *a priori* unknown. This is done by asserting that the compositions of the working gas correspond to $\Delta_{^{13}CH_3D} = 0$‰ and $\Delta_{CH_2D_2} = 0$‰. Based on the calculations shown in Table 4.2 these measurements are converted to the "thermodynamic" (or absolute) reference frame.

Experiments vs theory are consistent with a 1:1 line (dashed line) with respect to the temperature differences. Composition of the working gas can be constrained by interpolation (e.g., where the 1:1 dashed line intersects the x axis) for the use of re-porting measurements in the thermodynamic reference frame (absolute) represented

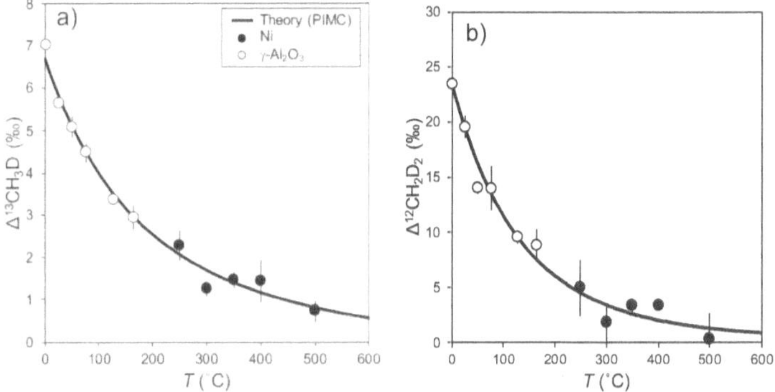


Figure 4.4: **Temperature dependence of the Δ values in thermodynamic (absolute) reference frame.** Note that experimental Δ values are based on converting $\Delta_{wg}$ (reported in the "working gas reference frame") into Δ values in the "thermodynamic reference frame."

by the PIMC calculations.

Fig. 4.3 compares the measured vs calculated clumped isotope compositions computed with PIMC at the experimental temperatures. We compare experimentally measured values as $1000 \times \ln\left(\Delta_{(wg)}/1000 + 1\right)$ values vs. computed values of $1000 \times \ln(K_{^{13}\text{CH}_3\text{D}})$ (panel a) and $1000 \times \ln\left(8/3 \times K_{\text{CH}_2\text{D}_2}\right)$ (panel b) at the experimental temperatures. We note that the measured Δ-values (y-axis) are plotted as $1000 \times \ln\left(\Delta_{wg}/1000 + 1\right)$ since the measured values follow the $1000 \times (R/R* -1)$ notation, while theory appropriately follows the $1000 \times \ln(K)$ notation[91], and by converting the measured $1000 \times (R/R*-1)$ values to $1000 \times \ln(R/R*)$ values we place the measured and computed values on more comparative grounds. Since experimentally measured values are shown in the working gas reference frame, the key aspect of this comparison is the relative difference between theory and experiment as a function of temperature. A least squares linear regression through each measured vs. theory (PIMC) dataset yields a slope of $1.02 \pm 0.04$ for the $\Delta_{^{13}\text{CH}_3\text{D}}$ comparison and $0.98 \pm 0.05$ for the $\Delta_{\text{CH}_2\text{D}_2}$ comparison (1 s.e.). Thus, both slopes are within 1 s.e. error of 1 over a temperature range of 1-500°C. Given this 1:1 agreement between experiment and theory, lines with slopes of 1 are used to infer the intercept in each plot to obtain estimates of the working gas composition. These yield $\Delta_{^{13}\text{CH}_3\text{D}} = 2.59 \pm 0.14‰$ and $\Delta_{\text{CH}_2\text{D}_2} = 5.86 \pm 0.60‰$ (1 s.e.). As expected from the 1:1 agreement in Fig. 4.3, the experimental data match the predicted tem-

---

[a]computed from Eqs. (4.5) and (4.6)

[b]To facilitate a more direct comparison, these values are reported as $1000 \times \ln(\Delta/1000 + 1)$.

[c]In this instance, error is 1 s.e. (internal precision) because the value represents a single measurement.

| T (°C) | PIMC $\Delta^{eq}_{^{13}CH_3D}$ [a] | Experiment $\Delta_{^{13}CH_3D}$ [b] | $1\sigma$ | residuals | PIMC $\Delta^{eq}_{CH_2D_2}$ [a] | Experiment $\Delta_{CH_2D_2}$ [b] | $1\sigma$ | residuals |
|---|---|---|---|---|---|---|---|---|
| 500 | 0.82 | 0.76 | 0.23 | -0.06 | 1.30 | -0.06 | 1.93 | -1.36 |
| 400 | 1.18 | 1.44 | 0.51 | 0.27 | 2.03 | 3.45 | 0.40 | 1.42 |
| 350 | 1.41 | 1.46 | 0.17 | 0.05 | 2.60 | 3.43 | 1.21[c] | 0.83 |
| 300 | 1.70 | 1.26 | 0.16 | -0.45 | 3.39 | 1.86 | 1.72 | -1.52 |
| 250 | 2.07 | 2.27 | 0.34 | 0.20 | 4.47 | 4.99 | 2.49 | 0.52 |
| 165.4 | 2.95 | 2.93 | 0.30 | -0.02 | 7.43 | 8.84 | 1.44 | 1.42 |
| 127.8 | 3.50 | 3.35 | 0.11 | -0.14 | 9.46 | 9.59 | 0.59 | 0.14 |
| 75.7 | 4.48 | 4.47 | 0.21 | -0.01 | 13.45 | 13.95 | 1.98 | 0.50 |
| 50.5 | 5.09 | 5.07 | 0.24 | -0.01 | 16.08 | 13.99 | 0.47 | -2.09 |
| 25 | 5.81 | 5.62 | 0.07 | -0.19 | 19.36 | 19.45 | 0.95 | 0.08 |
| 1.2 | 6.64 | 7.00 | 0.09 | 0.36 | 23.15 | 23.23 | 1.55 | 0.08 |
| $1\sigma$ | | | | **0.22** | | | | **1.17** |

Table 4.4: Comparison between theoretical (PIMC) and experimental $\Delta$ values in ascending order (highest temperature first).

perature dependence from the PIMC calculations over 1-500°C (see Fig. 4.4 and Table 4.4). The computed $\pm 1\sigma$ of the residuals (Table 4.4) are $\pm 0.22‰$ and $\pm 1.17‰$ for the $\Delta_{^{13}CH_3D}$ and $\Delta_{CH_2D_2}$ values, respectively, and are comparable to the external precision estimated solely from the experimental replicates at a given temperature ($\pm 0.28‰$ for $\Delta_{^{13}CH_3D}$ and $\pm 1.61‰$ for $\Delta_{CH_2D_2}$, $1\sigma$).

This work yields the important and satisfying result that theoretically calculated $\Delta^{eq}_{^{13}CH_3D}$ and $\Delta^{eq}_{CH_2D_2}$ values using the most rigorous theoretical approach available (PIMC) are in 1:1 (at the 1.s.e. level) agreement with experimental determinations of equilibrium $\Delta^{eq}_{^{13}CH_3D}$ and $\Delta^{eq}_{CH_2D_2}$. This provides confidence in both the theory, experiments, and measurement techniques over essentially the full range of formation temperatures of microbial and thermogenic gases on Earth.

Finally, the working gas clumped compositions yield apparent methane-clumped isotope temperatures of $196 \pm 13°C$ for $\Delta_{^{13}CH_3D}$ and $204 \pm 17°C$ for $\Delta_{CH_2D_2}$ (1 s.e.) as determined by the polynomial fit to the PIMC calibration (Eqs. 4.7 & 4.8). Based on the $\delta D_{VSMOW}$ ($-159.3 \pm 2.4‰$) and $\delta^{13}C_{VPDB}$ ($-38.37 \pm 0.2‰$) values of the working gas, the cylinder gas is likely thermogenic in origin.[107,108] Such temperatures are reasonable potential gas formation temperatures[128] and are consistent with the common observation that apparent methane clumped isotope temperatures of thermogenic methane are compatible with expectations of thermogenic gas formation temperatures.[113,116,119,120]

The $\Delta_{^{13}CH_3D}$ and $\Delta_{CH_2D_2}$-based temperatures are within 1 s.e. of each other. Such agreement has been previously seen both in assumed thermogenic gases from commercially purchased cylinders[125] as well as thermogenic gases from natural gas deposits.[113,116] Such agreement has been taken as additional evidence that thermogenic gases may form in clumped-isotope equilibrium and that $\Delta_{^{13}CH_3D}$ and $\Delta_{CH_2D_2}$ may represent formation temperatures of thermogenic gases (or at least re-equilibration temperatures).

Given the agreement in clumped-based temperatures of our working gas inferred for both $\Delta_{^{13}CH_3D}$ and $\Delta_{CH_2D_2}$, we could choose to force our working gas to have a $\Delta_{CH_2D_2}$ composition that corresponds to the temperature derived from the $\Delta_{^{13}CH_3D}$ calibration ($\sim196°C$) given that the $\Delta_{^{13}CH_3D}$ measurements are more precise. This exercise would yield a $\Delta_{CH_2D_2}$ value of $\sim6.17‰$ for our working gas using Eq. (4.8), which is about 0.31‰ higher than, but within 1 s.e. of what we directly infer from our calibration (5.86 ± 0.60‰, 1 s.e.). This may mean that our future measurements of $\Delta_{CH_2D_2}$ could be biased to ca. 0.3‰ higher values based on our calibration, but given our typical external precision ($\sim1.4‰$) we do not expect that any such bias would change any interpretations of environmental or experimental samples.

**Comparison of the PIMC calculations to the BMU approximation**

PIMC calculations provide a means to compute stable isotope fractionation factors independent of the traditionally employed BMU model and are more rigorous and accurate (assuming a high-quality PES and sufficiently large number of beads $n$ and number of samples for the systematic and statistical error convergence, respectively). Therefore, comparison of BMU and PIMC calculations can be used to identify errors in BMU calculations.[86,93] In the current study, all BMU-computed RPFRs exhibit significant departures from the PIMC-computed RPFRs: up to 5-6‰ for $^{13}CH_4/CH_4$ & $^{13}CH_3D/CH_3D$ RPFRs and 29-95‰ for the $CH_3D/CH_4$ & $CH_2D_2/CH_3D$ RPFRs over the computed temperature range (-3 to 527°C). Given that both BMU and PIMC computations were performed using the same PES for methane computed using high level coupled cluster theory,[133] these are true differences between the BMU and PIMC theoretical treatments. The $CH_3D/CH_4$ RPFR exhibit 5-20x larger relative errors than $^{13}CH_4/CH_4$ RPFRs, where the larger discrepancy present for D/H exchange is likely due to the long-recognized inadequacies in the simplified treatments of partition functions (PF) in the BMU model to account properly for D/H exchange (harmonic vibrational PF, classical rotational PF).[90,92,96,132] The PIMC calculations inherently account for vibrational anharmonicity and quantize the rotational motions, and therefore avoid these well-understood pitfalls of the BMU approach.

Additional insight into the problem may be given by comparisons between the calculations of the present study (BMU vs. PIMC) and previous BMU calculations performed with and without anharmonic corrections reported by Liu and Liu[130]. We first note that such a comparison is ultimately inexact because their[130] calculations are based on electronic potential energy surfaces for methane computed at the MP2 level of theory rather than the more accurate couple cluster theories of the present study.[133] Nevertheless, the relative difference between uncorrected and corrected $CH_3D/CH_4$ RPFRs using the BMU model by Liu and Liu (ca. 112 to 39‰ over -3 to 527°C, respectively, is comparable to what we observe for BMU and PIMC calculations in this study (95-29‰ over -3 to 527°C, respectively). The six corrections applied include those accounting for vibrational anharmonicity and quantum corrections to rotational motions among others.[130,132] The total correction given by Liu and Liu (a multiplicative factor of ~0.90 to ~0.96 from -3°C to 527°C) is almost entirely driven by the correction for the anharmonic contributions to the zero point energy. This may tentatively suggest that the harmonic treatment of the vibrational partition function may be the source of much of the error in BMU-based D/H-related computations for methane.

Regardless of the precise source of the errors in the BMU model, the contrastingly small ($\leq$0.1-0.4‰) relative differences in the computed equilibrium constants and related $\Delta$ values describing equilibrium clumping in methane from BMU-RPFRs arises from a cancellation of errors in component RPFRs as observed by Webb and Miller.[86] One likely reason for this precise cancelation of errors may be due to inherent symmetry preserved in these isotopic clumping reactions. In particular, any errors present in the $^{13}CH_4/CH_4$ RPFR are expected to be similar in nature and magnitude to those present in the $^{13}CH_3D/CH_3D$ RPFR, since the RPFRs reflect the same type of isotopic substitution. The same cannot be said for some exchange reactions involving isotopomers (e.g., $^{14}N^{15}NO \rightleftharpoons {}^{15}N^{14}NO$) for which BMU calculations have been shown to only benefit from a partial cancelation of errors.[86] Although the RPFR errors are significant on the per mil scale, we find it remarkable that such errors only amount to relative free energy differences of approximately $10^{-3}$ and $2 \times 10^{-2}$ kcal/mol for the $^{13}C/C$-related and D/H-related RPFRs, respectively.

## 4.5  Summary

We presented a new experimental and theoretical working gas calibration method, covering the range of expected thermogenic and microbial gas formation temperatures on Earth. On the experimental side we utilize $\gamma$-$Al_2O_3$ an Ni catalysts to allow for methane equlibration from 1-500°C, and on the theoretical side the path integral calculations (PIMC) of equilibrium clumping in methane over the same temperature range. We observed 1:1 agreement between measured differences in $\Delta_{^{13}CH_3D}$ and $\Delta_{CH_2D_2}$ for samples equilibrated from 1-500°C vs. theoretical (PIMC) calculations of $\Delta_{^{13}CH_3D}$ and $\Delta_{CH_2D_2}$ over the same temperature range. We used the PIMC calculations to gain insight into the relative source of errors in the approximate BMU approach.

# Chapter 5

# HYDROGEN ISOTOPIC EQUILIBRIUM IN THE SYSTEM $CH_4-H_2-H_2O$

## 5.1 Introduction

Stable carbon and hydrogen isotopic compositions are commonly used to trace the source and sinks of methane ($CH_4$) in a variety of systems including economic hydro-carbon reservoirs, sediments, lakes, the ocean, hydrothermal systems, and volcanic systems. The basis of this approach is that methane formed by microbial, thermo-genic, and abiotic formational processes often (though not always) occupy different fields in plots of D/H (given by $\delta$D, Eq. 3.2) vs. $^{13}$C/C (given by $\delta^{13}$C).[107,108,134,135]

Stable isotopic composition of methane generated by a given process is controlled by (i) the source isotopic composition of the carbon and hydrogen[107,110,134,136] and (ii) the isotope effects of the chemical reactions involved in methane formation. The carbon and hydrogen isotopic compositions of thermogenic methane are com-monly thought to be controlled by kinetic isotope effects.[137] For microbial gases, both kinetic and equilibrium carbon and hydrogen isotope effects are thought to control methane's isotopic composition.[107,108,138,139] Finally, in high-temperature settings, such as volcanic and hydrothermal systems, equilibrium isotope effects between methane and water ($H_2O$) or methane and carbon dioxide ($CO_2$) have been proposed to set methane's isotopic composition,[140,141] though there is also an alter-native interpretation.[142] As such, equilibrium processes are commonly thought to be involved in setting the carbon and hydrogen isotopic composition of methane in nature. Here, we provide both experimental and theoretical calibrations of hydrogen isotopic equilibrium between methane, molecular hydrogen ($H_2$), and liquid water from 3-200°C. To place this work into context, we first review the history and current thinking on the role of equilibrium processes in setting the carbon and hydrogen isotopic composition of methane. Second, we review previous experimental and the-oretical calibrations of the temperature dependence of carbon isotopic equilibrium between methane vs. $CO_2$ and hydrogen isotopic equilibrium between methane vs. liquid water.

**The role of isotopic equilibrium in setting the isotopic composition of methane**

It is commonly assumed that kinetic processes and therefore kinetic isotope effects largely control the carbon and hydrogen isotopic composition of thermogenic and microbial methane. However, over the past 50 years there have been a series of proposals that equilibrium isotope effects also play a role. We begin with a review of previous work on carbon isotopes followed by hydrogen isotopes.

Murata et al. proposed that sedimentary methane and $CO_2$ can exchange carbon isotopes on geological timescales and come into carbon isotopic equilibrium in both low-temperature sedimentary systems and warmer natural gas reservoirs and hot springs.[143] This proposal was based on the observation that measured differences in $\delta^{13}C$ of $CH_4$ and $CO_2$ yield reasonable temperatures based on a sample's collection environment and the assumption that methane and $CO_2$ are in carbon isotopic equilibrium. Moreover, $\delta^{13}C$ values of sedimentary samples of microbial methane and $CO_2$ yield reasonable calculated temperatures assuming the two are in carbon isotopic equilibrium.[144] Finally, based on similar arguments, it was proposed that methane in Australian coal-seam gases also approaches carbon isotopic equilibrium with $CO_2$.[145] In settings where microorganisms might be expected to be active, this equilibrium is thought to be achieved through isotope-exchange reactions catalyzed by methanogenic microbial organisms.[146]

The hypothesis that methane and $CO_2$ can achieve carbon isotopic equilibrium in low-temperature sedimentary environments was originally largely rejected on the grounds that the rates of carbon isotope exchange are too sluggish to promote equilibration on geological timescales.[107,147,148] These arguments have led to the common assumption that carbon isotope effects for microbial methane generation are controlled by kinetic isotope effects.

Valentine et al. proposed that both equilibrium and kinetic isotope effects set the carbon isotopic composition of microbial methane.[138] Specifically, they proposed that the free energy available to drive microbial methane generation dictates the overall degree of reversibility of enzymes involved in the reduction of $CO_2$ to methane. When free energy gradients are low, enzymes catalyze reactions in both the forward and reverse direction (i.e., are reversible) and thus catalyze both the forward reduction of $CO_2$ to methane and the reverse oxidation of methane back to $CO_2$. Such reversibility allows for carbon isotopic equilibration to occur between $CH_4$ and $CO_2$. In contrast, when free energy gradients are high, enzymes act irreversibly and only catalyze the forward reduction of $CO_2$ to methane. Under these conditions, only kinetic isotope effects are expressed.

Recently, using a reaction-diffusion model, Meister et al. found that fractionation factors required to model the observed differences in the $\delta^{13}C$ of $CO_2$ and biogenic methane in marine sediments are consistent with what would be expected if the two forms were in isotopic equilibrium.[149] Based on this, they suggested that methanogens in deep-sea sediments could promote $CH_4 - CO_2$ carbon isotopic equilibration during methane generation and further argued that methanogens can produce methane in carbon isotopic equilibrium with $CO_2$ because the carbon isotopic composition of microbially generated methane reacted with $CO_2$ from experiments[150] are similar to the values expected for equilibrium at the correspond growth temperatures (35-85°C). Thus the problem has, over the past 50 years, come full circle, with the initial proposal that methanogens catalyze equilibration of carbon between methane vs. $CO_2$ in marine sediments, though initially dismissed, receiving renewed support.

Microbially catalyzed carbon isotopic equilibration between methane and $CO_2$ has also been proposed to occur during anaerobic oxidation of methane at the sulfate-methane transition zone.[151] Their suggestion is that enzymes of anaerobic methanotrophs can operate reversibly and thus catalyze both the forward oxidation of methane to $CO_2$ and the reduction of $CO_2$ back to methane with the degree of exchange a function of free energy available to the system.[151–153]

The carbon isotopic composition of thermogenic methane is generally thought to be controlled by kinetic isotope effects.[137] However, the thermal decomposition of acetic acid yields $CO_2$ and methane with carbon isotopic compositions consistent with generation in carbon isotopic equilibrium from 290 to 650°C.[154] Recently, based on observed differences between the $\delta^{13}C$ of $CO_2$ and thermogenic methane vs. measured methane clumped-isotope based temperatures, it was proposed that thermogenic methane and $CO_2$ can achieve carbon isotopic equilibrium in the subsurface through reactions that promote methane oxidation and $CO_2$ reduction.[155]

Equilibrium isotope effects have also been suggested to set the hydrogen isotopic composition of some microbial and thermogenic methane samples. For microbial gases, it has been observed that samples that yield methane clumped-isotope-based temperatures consistent with expected formation temperatures also yield differences between the $\delta D$ of methane and $H_2O$ that would be predicted if these samples formed in $CH_4 - H_2O$ hydrogen isotopic equilibrium.[119,121,123]. This pattern was explained as the result of high degrees of enzymatic reversibility during methanogenesis that catalyzes $CH_4 - H_2O$ hydrogen isotope-exchange reactions, equilibrating the two and promoting methane clumped-isotope equilibrium.[119,121] Based on examinations of biogenic methane and water hydrogen isotope systematics in coal and shale gas systems, it was proposed that microbial methane formed in these systems forms in hydrogen isotopic equilibrium with co-occurring waters.[156] Such patterns could also be explained by methanogens generating methane out of hydrogen isotopic equilibrium with water and then that microbes later catalyze hydrogen isotope-exchange reactions between methane and water to equilibrate the two.[139] Finally, it has been proposed that methane clumped isotopic equilibrium can be promoted during anaerobic methane oxidation.[113,116,117,121,157] In this explanation, anaerobic methane oxidizing archaea also operate enzymes sufficiently reversibly such that they promote $CH_4 - H_2O$ hydrogen isotope-exchange reactions.

It has also been proposed that the hydrogen isotopic composition of thermogenic methane can be influenced by equilibrium processes at elevated temperatures. More specifically, the suggestion based on relationships between $\delta^{13}C$ and $\delta D$ values of methane from Paleozoic deposits in the Appalachian Basin, is that methane begins exchanging and therefore equilibrating hydrogen isotopes with water at temperatures of 200 to 300°C.[158] Additionally, it was proposed that thermogenic methane may form in (or achieve) hydrogen isotopic equilibrium with water based on the observed agreement between apparent methane clumped-isotope temperatures (from 118-204°C) and measured differences in the $\delta D$ of methane and water versus those expected for $CH_4 - H_2O$ hydrogen isotopic equilibrium.[116] Recently, it was proposed that methane equilibrates hydrogen with other gaseous alkanes (e.g., ethane, propane, etc.) in thermogenic gas reservoirs. This proposal is based on the observation that measured methane clumped-isotope-based temperatures are similar to temperatures calculated based on the assumption of hydrogen isotopic equilibrium between methane and other alkanes.[155]

Finally, as noted above, high temperature (>275°C) volcanic and hydrothermal systems are commonly thought to yield methane in isotopic equilibrium with co-occurring water and $CO_2$,[118,140,141,144,159,160] although carbon isotopic compositions of methane and $CO_2$ can also be out of equilibrium as well.[142]

Evaluation of whether methane forms in or later achieves isotopic equilibrium with either $CO_2$ or water requires constraints on the equilibrium fractionation factors between methane and these gases and liquids at relevant environmental temperatures. All microbial and most thermogenic methane is thought to form at temperatures below 200°C.[128,161] In contrast, all experimentally determined equilibrium fractionation factors for $CH_4(g) - CO_2(g)$ carbon and $CH_4(g) - H_2O(l)$ hydrogen isotopic equilibrium exist only for temperatures greater than 200°C.[140,162,163] As a result, either experimental determinations of equilibrium fractionation factors must be extrapolated to lower temperatures or calculation used when studying all microbial and most thermogenic samples.

In the following section, we discuss the current knowledge of the equilibrium isotopic composition of methane vs. $CO_2$ for carbon and methane vs. liquid water for hydrogen based on experimental and theoretical studies.

**Carbon isotopic equilibrium between $CH_4$ and $CO_2$ and hydrogen isotopic equilibrium between $CH_4$ and liquid water**

Isotopic differences between two phases or species are given using the "alpha" notation, Eq. (3.3). We give values of $\alpha$ as $1000 \times \ln \alpha$ as this form has a theoretically based dependence on temperature for systems at isotopic equilibrium.[83]

The equilibrium carbon isotopic composition of $CH_4$ vs. $CO_2$ in the gas phase as a function of temperature has been determined experimentally in two studies: the first from 200 to 600°C and the second from 300 to 1200°C.[162,163] These studies yield equilibrium $1000 \times \ln {}^{13}\alpha_{CH_4(g)-CO_2(g)}$ values in agreement at overlapping experimental temperatures (within 0.01 to 1.01‰ from 300 to 600°C). Additionally, theoretical and experimental estimates of $1000 \times \ln {}^{13}\alpha_{CH_4(g)-CO_2(g)}$ for $CH_4$ – $CO_2$ carbon isotopic equilibrium agree at overlapping temperatures: theoretical predictions[90] are offset by 0.89‰ from experimental data of Horita et al..[162] The combination of both experimental studies are offset from the calculation[90,144] by 0.2 to 0.6‰ and 0.7-1.2‰ respectively.

Both studies do not recommend extrapolation of their results to temperatures outside of their calibrated range due to their use of polynomial and power-law fit, respectively.[162,163] Instead, both studies recommend that if extrapolation is needed, that the temperature dependence of calculation be fitted to the experimental data and that these fits be used for any extrapolations beyond the experimentally calibrated temperature range.[164] In such an approach, it is not the absolute values of $1000 \times \ln \alpha$ from the theoretical studies that matter, but rather the predicted change in $1000 \times \ln \alpha$ as a function of temperature (i.e., the temperature dependence) as any constant offsets between theoretical and experimental studies will be minimized during fitting of the calculation to the experimental data.

We are aware of six published theoretical estimates of $1000 \times \ln {}^{13}\alpha_{CH_4(g)-CO_2(g)}$ as a function of temperature at isotopic equilibrium that could be used for such an exercise.[90,144,155,159,165,166] For temperatures lower than those accessed by experiments (i.e., below 200°C), five of them are in general agreement with maximum differences of 2.3‰ between calculations from 0 to 200°C. Over this temperature range, these studies yield similar temperature dependencies: calculated differences in $1000 \times \ln {}^{13}\alpha_{CH_4(g)-CO_2(g)}$ range from 44.1‰ to 45.2‰ at 0 vs. 200°C. Values from Craig (1953) from 0 to 200°C are offset from other theoretical studies by up to 12‰, with predicted change in value over this temperature range of 39.6‰ (indicating a different temperature dependence as well).[159] This difference was attributed to improvement in the accuracy of spectroscopic data from the 1950s to 1970s.[162] Regardless, the strong agreement between post 1950s calculation of $1000 \times \ln {}^{13}\alpha_{CH_4(g)-CO_2(g)}$ and agreement with experimental determinations at overlapping temperatures provides confidence in using theory as a basis to extrapolate experimental calibrations of $1000 \times \ln {}^{13}\alpha_{CH_4(g)-CO_2(g)}$ to temperatures below the experimentally calibrated range (<200°C).

An experimental determination of equilibrium ${}^{D}\alpha_{CH_4(g)-H_2O(l)}$ exists for temperatures from 200 to 500°C.[140] This is not a direct determination based on co-equilibration of $CH_4(g)$ and $H_2O(l)$; instead, they first equilibrated the hydrogen isotopes of $CH_4$ and $H_2$ gas using nickel-thoria catalysts from 200 to 500°C and derived a calibration for ${}^{D}\alpha_{CH_4(g)-H_2O(l)}$ vs. temperature. They then combined this expression with other experimentally based estimates of equilibrium ${}^{D}\alpha_{H_2O(l)-H_2(g)}$ values to derive an equation for ${}^{D}\alpha_{CH_4(g)-H_2O(l)}$ vs. temperature, noting that further experiments were needed to constrain the temperature dependence of ${}^{D}\alpha_{CH_4(g)-H_2(g)}$

and $^{D}\alpha_{CH_4(g)-H_2O(l)}$ at temperatures below their experimental range (<200°C).

Beyond the typical concerns of extrapolating experimentally derived equilibrium $\alpha$ values outside of their calibrated range, the specific extrapolation of this calibration of $^{D}\alpha_{CH_4(g)-H_2O(l)}$ to temperatures below 200°C has additional complexity. They provide a fit to their data with $^{D}\alpha_{CH_4(g)-H_2O(l)}$ linearly dependent on $\frac{1}{T^2}$. However, such a dependence for $^{D}\alpha_{CH_4(g)-H_2(g)}$ is inconsistent with the calculations.[90] Instead, it is $\ln\,^{D}\alpha_{CH_4(g)-H_2(g)}$ that has a theoretically based dependence on $1/T$. As such, extrapolation of this calibration could lead to inaccurate values of $^{D}\alpha_{CH_4(g)-H_2(g)}$ at temperatures below 200°C.

One way around this would be to use the equilibrium theoretical estimates of $1000 \times \ln\,^{D}\alpha_{CH_4(g)-H_2O(l)}$ vs. temperature are fit to experimental data and then used as the basis for the extrapolation.[164] This has not been attempted for this system. The question is whether, as was seen above for the case of carbon isotope equilibrium between $CO_2$ and methane, current calculation are in agreement with the experimental results at overlapping temperatures and if there is general agreement in the temperature dependence of $1000 \times \ln\,^{D}\alpha_{CH_4(g)-H_2(g)}$ at temperatures below those calibrated experimentally.

To our knowledge, no calculation of $1000 \times \ln\,^{D}\alpha_{CH_4(g)-H_2O(l)}$ vs. temperature at isotopic equilibrium exist. Instead, calculation for hydrogen isotopic equilibrium between gaseous methane and water (i.e., $1000 \times \ln\,^{D}\alpha_{CH_4(g)-H_2O(g)}$ exist and can be converted to $1000 \times \ln\,^{D}\alpha_{CH_4(g)-H_2O(l)}$ using the experimental calibration of $1000 \times \ln\,^{D}\alpha_{H_2O(l)-H_2O(g)}$.[167] We use the experimental (and not theoretical) gas-liquid water calibration because it was used for higher temperatures,[140] making comparison of theory to their experimental calibration consistent. We are aware of seven distinct theoretical estimates from four studies for $1000 \times \ln\,^{D}\alpha_{CH_4(g)-H_2O(g)}$ vs. temperature.[90,130,144,166] In all studies, calculations were done at a minimum temperature of 0°C and maximum temperatures equal to or greater than 500°C. These theoretical studies disagree significantly from 0 to 200°C on the value of $1000 \times \ln\,^{D}\alpha_{CH_4(g)-H_2O(l)}$ at isotopic equilibrium (Fig. 5.1). For example, at 0°C there is a maximum disagreement of up to 159‰ between the theoretical studies and up to 164‰ between the theoretical studies and the extrapolation of Horibe and Craig's calibration. This disagreement is also seen in the calculated temperature dependence over this range (0 to 200°C): with differences up to 67‰ between theoretical studies and up to 91‰ between theory and the extrapolation of Horibe and Craig's calibration. Thus, there is significant uncertainty (order 100‰) both on the correct absolute values and temperature dependence of $1000 \times \ln\,^{D}\alpha_{CH_4(g)-H_2O(g)}$ from 0 to 200°C. This uncertainty makes it challenging to confidently extrapolate the calibration of Horibe and Craig to low (<200°C) temperatures based on theory.

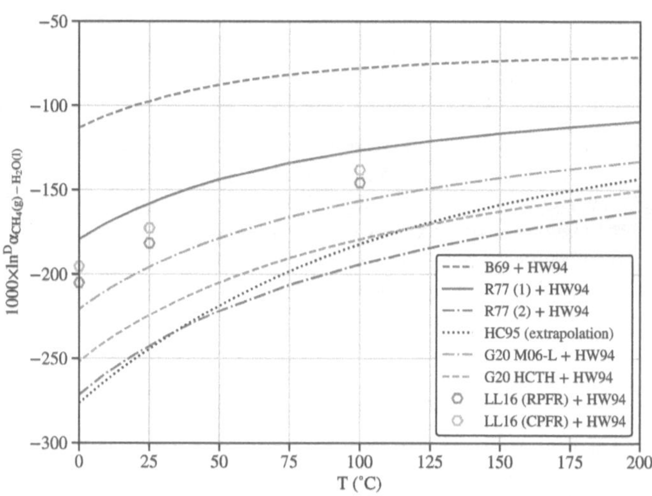



Figure 5.1: **Published estimates of equilibrium** $1000 \times \ln^D \alpha_{CH_4(g)-H_2O(l)}$ from 0 to 200°C. Theoretical estimates are based on calculations of $1000 \times \ln^D \alpha_{CH_4(g)-H_2O(g)}$ that are converted to $1000 \times \ln^D \alpha_{CH_4(g)-H_2O(l)}$ using the experimental calibration of $1000 \times \ln^D \alpha_{H_2O(l)-H_2O(g)}$ from [167] — this is noted as "+ HW94" when applied. HC95 (extrapolation) is the extrapolation of the equation given in Ref. [140] to temperatures below the range of their experiments (200 to 500°C). B69 is Ref. [144], R77 is Ref. [90], where (1) refers to the results where the excess factor $X_{CH_4}$ was calculated using the harmonic approximation, while R77 (2) refers to the use of an anharmonic approximation in this calculation. LL16 is Ref. [130] using their notations where "RPFR" are calculations performed assuming a harmonic oscillator approximation while "CPFR" (to use their terminology) includes higher order corrections for anharmonicity, rotational-vibrational coupling, and other terms. G20 M06-L is M06-L calculated RPFR as presented in [166], while G20 HCTH is HCTH calculated RPFR from the same work. B69, R77 (1), and R77 (2) are 4th order polynomial fits of $1000 \ln^D \alpha_{CH_4(g)-H_2(g)}$ vs. $1/T$ based on calculated values of RPFRs from the given study.

56

**This study**

Here, we provide an experimentally based calibration of the equilibrium hydrogen isotope fractionation factor for methane and $H_2$ from 3 to 200°C. Based on previous experimental determinations of hydrogen isotope equilibrium between $H_2$, $H_2O(g)$, and $H_2O(l)$, we provide an interpolatable calibration of $1000 \times \ln {}^D\alpha_{CH_4(g)-H_2O(l)}$ derived from experimental constraints from 3 to 200°C. We additionally provide new theoretical estimates for hydrogen isotopes for $1000 \times \ln {}^D\alpha^{eq}_{CH_4(g)-H_2(g)}$, $1000 \times \ln {}^D\alpha^{eq}_{CH_4(g)-H_2O(g)}$, and $1000 \times \ln {}^D\alpha^{eq}_{H_2O(g)-H_2(g)}$ and carbon isotopes for $1000 \times \ln {}^{13}\alpha^{eq}_{CH_4(g)-CO_2(g)}$ based on the PIMC calculations. We compare these estimates to our and other experimentally determined calibrations.

## 5.2 Methods

The experimental procedures are detailed in [99] — we only describe the theoretical methodology here. calculation for isotopic equilibrium were based on calculations of reduced partition function ratios for isotopologues with one rare isotope vs. the unsubstituted molecule as described in section3.2. We calculated the RPFRs using two distinct approaches: (i) using the BMU (harmonic) approach[92,96] and (ii) using the PIMC approach.[3,86,102] In the harmonic approach, the total partition function is assumed to factorize into vibrational, rotational and translational components; then the vibrations are approximated as harmonic, rotations as rigid, and both rotations and translations are assumed to be classical. The PIMC approach includes a fully anharmonic and quantum mechanical description of the RPFRs and is thus a more rigorous theoretical treatment as compared to harmonic calculations. Indeed, PIMC calculations have been used to identify sources of error in harmonic calculations of RPFRs.[86,93]

When computing RPFRs we neglected the effects of: (i) intermolecular interactions, since the gases are dilute; (ii) electronic excitations, since the excited states are well-separated for all the molecules considered here; (iii) internal structure of the nuclei; and (iv) relativistic effects, since the molecules only contain light atoms.[168]

**Potential energy surfaces and harmonic frequencies**

In order to calculate an RPFR using either the harmonic or PIMC approach, the PES of the molecule must first b e s pecified. Fo r th e PI MC ca lculations, we used published potentials for $CH_4$, $CO_2$, and $H_2O$, and calculated ourselves the surface for $H_2$. These potential energy surfaces are referred to as "reference" potentials. The potential for methane was taken from Ref. [133], where it was calculated at the coupled-cluster level with single double excitations with triple excitations included perturbatively (i.e., CCSD(T)) using correlation-consistent polarized triple zeta (cc-pVTZ) and quadruple zeta (cc-pVQZ) basis sets. The PES for $H_2O(g)$ was taken from [169], where it was computed at internally contracted multireference configuration i nteraction ( icMRCI) a nd C CSD(T) l evels o f theory with augmented correlation-consistent polarized quintuple-zeta (aug-cc-pV5Z) and correlation-consistent polarized sextuple zeta (cc-pV6Z) basis sets and three-body terms fitted to reproduce the experimental line positions in rovibrational spectra of water. The carbon dioxide reference potential is a CCSD(T)/aug-cc-pVTZ surface refined based on extrapolation to the one-particle basis set limit, corrections for scalar relativity, higher-order electron correlations, and spectroscopic data from the HITRAN2008 database from Ref. [170].

We calculated the PES for $H_2$ using the Molpro software package (version 2019.2) with CCSD (exact for the two-electron problem) and cc-pVQZ basis set. The one-dimensional surface is obtained through spline interpolation between a dense set of single point CCSD/cc-pVQZ energy calculations between 0.38 and 1.8 Å with the interval of 0.005 Å. We tighten the energy and orbital convergence thresholds to $10^{-16}$ and leave the other input parameters on standard settings. We have also confirmed that larger basis sets (up to aug-cc-pV6Z) do not significantly change hydrogen's vibrational frequency (within $\pm 3$ cm$^{-1}$). Energy outside the computed range is approximated by the following fit:

$$E(q) = \begin{cases} E_{eq} + \exp\left\{-9.56(q - q_0)\right\}, & q < 0.38\text{Å} \\ E_\infty - \exp\left\{-1.85q\right\}, & q > 1.8\text{Å} \end{cases} \tag{5.1}$$

where energy is in units of Hartree, $E_{eq} = -1.17379647$ is the equilibrium (minimum) energy, $E_\infty = -1$ is the energy of two hydrogen atoms at infinite distance and $q_0 = 0.23$ Å. The high vibrational frequency of hydrogen ensures that the molecule only explores a tight range of molecular configurations around the equilibrium geometry, so we do not expect the accuracy of the fit above to influence computed RPFRs. Harmonic calculations of RPFRs require only the determination of the

| Molecule | $\omega_1$ | $\omega_2$ | $\omega_3$ | $\omega_4$ | $\omega_5$ | $\omega_6$ | $\omega_7$ | $\omega_8$ | $\omega_9$ |
|---|---|---|---|---|---|---|---|---|---|
| $CH_4$, [133] | 1345.3 | 1345.3 | 1345.3 | 1570.4 | 1570.4 | 3036.2 | 3157 | 3157 | 3157 |
| $^{13}CH_4$ | 1337 | 1337 | 1337 | 1570.4 | 1570.4 | 3036.2 | 3145.9 | 3145.9 | 3145.9 |
| $CH_3D$ | 1188 | 1188.09 | 1339.8 | 1508.1 | 1508.1 | 2285.2 | 3071.4 | 3156.8 | 3156.8 |
| $CO_2$, [170] | 672.8 | 672.8 | 1353.5 | 2395.9 | | | | | |
| $^{13}CO_2$ | 653.7 | 653.7 | 1353.5 | 2327.8 | | | | | |
| $H_2O$, [169] | 1649.1 | 3832.7 | 3944.3 | 1649.1 | | | | | |
| HDO | 1445.6 | 2824.3 | 3890.8 | 1445.6 | | | | | |
| $H_2$ | 4403.4 | | | | | | | | |
| HD | 3814 | | | | | | | | |

Table 5.1: Reference harmonic frequencies used to compute reference harmonic RPFRs.

harmonic frequencies at an energy-minimized geometry. The numerical Hessian is computed with 5- and 9-point stencil (in one and two dimensions, respectively) around known minimum on the reference potential energy surfaces. We converged the frequencies to 0.1 cm$^{-1}$. We calculated harmonic frequencies (see Table 5.1), obtained from the same reference potentials as used for the PIMC calculations. As such, these RPFRs can be compared directly to PIMC results and used to quantify the importance of anharmonic and quantum effects that are absent in the harmonic treatment; we later refer to these harmonic RPFRs as "reference harmonic" lines.

We also computed the harmonic frequencies from a variety of molecular structures optimized using a hierarchy of levels of theory; these included the restricted Hartree-Fock (RHF) method (a mean-field theory that does not take into account the electron-electron correlations) along with three successively better approximations for the correlation energy: second order Møller-Plesset perturbation theory (MP2) and coupled-cluster with single and double excitations (CCSD), as well as CCSD(T), where triple excitations are included perturbatively. These levels of theory were paired with basis sets of various sizes (cc-pVXZ and aug-cc-pVXZ as defined above, where X=D [double], T [triple], or Q [quadruple])[171] using the Molpro software package with the default settings. These calculations were done in order to determine the sensitivity of our calculation of RPFRs to electronic correlations and basis set completeness.

The RPFR of an isotopologue pair calculated using the harmonic approach is given by Eq. (3.25) PIMC methodology is described in section. 3.3 The direct scaled-coordinate estimator[102] was used to calculate the RPFR's for heavy vs. light isotopologues. Heavy isotopologue configurations were sampled with PIMC in Cartesian coordinates with an explicit staging transformation.[31] The staging length, $j$, was set such that 38-42% of all proposed staging moves are accepted. Prior to any data collection, each sampling trajectory was equilibrated for $10^5$ Monte Carlo steps, with $P/j$ staging moves (rounded up to the nearest integer) attempted per Monte Carlo step. Thereafter, ring-polymer configurations were sampled every 10 Monte Carlo steps. The total number of Monte Carlo moves for each partition function ratio calculation was $2 \times 10^8$.

Aside from neglecting nuclear exchange, PIMC calculations give an exact answer for RPFRs for a specified PES in the limit of infinite sampling and infinite number of beads n. In practice, a finite number of beads can be chosen to achieve target accuracy, while the number of samples controls statistical uncertainty. The number of beads employed in the PIMC calculations was determined based on explicit convergence tests for the individual RPFRs over the range of temperatures studied. We ensured that the accuracy was within the standard error of the mean. This error is reported for every PIMC calculation as a measure of statistical uncertainty.

Non-Born-Oppenheimer effects (i.e., inaccuracies associated with the use of the Born-Oppenheimer approximation) are usually negligible compared to the those inherent to the PES. However, diagonal Born-Oppenheimer corrections (DBO corrections) can become important in high-accuracy electronic structure calculations for small molecules.[172,173] DBO corrections are lowest order perturbation-theory corrections to the Born-Oppenheimer approximation that correct for the dependence of the electronic wave function on the nuclear coordinates when calculating the nuclear kinetic-energy contribution. We used the DBO corrections calculated[174] at the CCSD level with aug-cc-pCVTZ (the augmented core-valence) basis set[175,176] for molecules in optimized geometries. DBO corrections were assumed to be locally independent of the nuclear coordinates based on weak ($<5$ cm$^{-1}$) dependence for hydrogen around equilibrium.[177,178] With this assumption, the calculated energy shifts associated with this correction affect the RPFRs via a free energy shift according to Eq. (6.1)

While DBO corrections can become important for fractionation of isotopes of hydrogen between different chemical species, they vanish exactly[179] for self-exchange reactions, i.e., exchange reactions considered in[86,93,98,180] (where all reactants and products are isotopologues). Moreover, they can be neglected for heavy-atom fractionation processes, since they decrease rapidly with increasing mass.

## 5.3 Results

| t (°C) | CH$_4$ − H$_2$ | ±1 s.e.[a] | H$_2$O − H$_2$ | ±1 s.e. | CH$_4$ − H$_2$O | ±1 s.e. |
| --- | --- | --- | --- | --- | --- | --- |
| 0.98 | 1288.4 | 0.32 | 1395.2 | 0.34 | -106.7 | 0.33 |
| 2.94 | 1275.3 | 0.37 | 1382.9 | 0.41 | -107.6 | 0.25 |
| 9.9 | 1232.2 | 0.51 | 1341.2 | 0.48 | -109.0 | 0.32 |
| 14.57 | 1203.6 | 0.38 | 1313.6 | 0.37 | -110.0 | 0.22 |
| 25.4 | 1143.1 | 0.25 | 1256.1 | 0.27 | -113.0 | 0.22 |
| 35.4 | 1091.1 | 0.31 | 1205.3 | 0.30 | -114.3 | 0.29 |
| 50.9 | 1016.0 | 0.26 | 1133.5 | 0.30 | -117.4 | 0.27 |
| 75.65 | 911.4 | 0.28 | 1032.0 | 0.28 | -120.6 | 0.23 |
| 99.9 | 823.0 | 0.23 | 945.7 | 0.22 | -122.7 | 0.25 |
| 127.4 | 736.6 | 0.22 | 860.8 | 0.19 | -124.3 | 0.19 |
| 157.35 | 656.3 | 0.19 | 780.9 | 0.21 | -124.7 | 0.19 |
| 180 | 603.1 | 0.18 | 727.9 | 0.23 | -124.8 | 0.21 |
| 200.2 | 560.7 | 0.16 | 685.1 | 0.18 | -124.4 | 0.16 |
| 203.35 | 554.4 | 0.16 | 678.6 | 0.16 | -124.2 | 0.14 |
| 219.88 | 522.9 | 0.15 | 646.9 | 0.18 | -124.1 | 0.15 |
| 264.07 | 450.2 | 0.12 | 571.8 | 0.13 | -121.6 | 0.13 |
| 302.55 | 397.0 | 0.14 | 516.2 | 0.16 | -119.2 | 0.13 |
| 364.58 | 327.3 | 0.10 | 441.8 | 0.13 | -114.6 | 0.13 |
| 407 | 288.6 | 0.11 | 399.4 | 0.14 | -110.8 | 0.14 |
| 458.1 | 249.9 | 0.11 | 356.0 | 0.12 | -106.1 | 0.13 |
| 476.5 | 237.4 | 0.09 | 341.6 | 0.10 | -104.2 | 0.10 |
| 503.4 | 220.6 | 0.08 | 322.3 | 0.10 | -101.7 | 0.10 |

Table 5.2: DBO-corrected PIMC calculations of $1000 \times \ln {}^D\alpha^{eq}$ for hydrogen isotopic equilibrium between gaseous molecules.

---

[a]Standard error of the mean for each PIMC calculation is reported here as a measure of statistical uncertainty only.

| t (°C) | $CH_4(g) - CO_2(g)$ | ±1 s.e.[a] |
|---|---|---|
| 1.85 | -77.13 | 0.06 |
| 11.85 | -73.09 | 0.06 |
| 26.85 | -67.87 | 0.05 |
| 51.85 | -60.30 | 0.05 |
| 76.85 | -53.92 | 0.05 |
| 101.85 | -48.60 | 0.04 |
| 126.85 | -44.00 | 0.04 |
| 176.85 | -36.59 | 0.03 |
| 226.85 | -30.92 | 0.03 |
| 276.85 | -26.42 | 0.01 |
| 326.85 | -22.82 | 0.01 |
| 376.85 | -19.87 | 0.01 |
| 426.85 | -17.42 | 0.01 |
| 476.85 | -15.39 | 0.01 |
| 526.85 | -13.67 | 0.01 |
| 601.85 | -11.58 | 0.01 |
| 801.85 | -7.83 | 0.01 |
| 1001.85 | -5.62 | 0.01 |
| 1301.85 | -3.70 | 0.02 |

Table 5.3: DBO-corrected PIMC calculations of $1000 \times \ln {}^{13}\alpha^{eq}$ for carbon isotopic equilibrium.

The results of PIMC calculations for hydrogen-isotope RPFRs vs. temperature for $CH_4$, $H_2$, and $H_2O$ are given in Table 5.2 and the carbon-isotope RPFRs vs. temperature for $CH_4$ and $CO_2$ in Table 5.3. In these tables, all given PIMC values include DBO corrections; the differences between DBO-corrected and uncorrected values are given in Tables 5.4. For hydrogen-isotope calculations, DBO corrections are between -34 to -12‰ for $1000 \times \ln {}^{D}\alpha^{eq}_{H_2O(g)-H_2(g)}$, +12 to +34‰ for $1000 \times \ln {}^{D}\alpha^{eq}_{CH_4(g)-H_2O(g)}$, and +0.1 to +0.3‰ for $1000 \times \ln {}^{D}\alpha^{eq}_{CH_4(g)-H_2(g)}$ from 0 to 500°C. For carbon $1000 \times \ln {}^{13}\alpha^{eq}_{CH_4(g)-CO_2(g)}$ calculations, DBO corrections are between -0.3 and -0.05‰ from 0 to 1300°C. For consistency, we use the DBO-corrected PIMC values in all cases, even when the correction is minor.

Values of $1000 \times \ln {}^{D}\alpha^{eq}$ for $CH_4(g) - H_2(g)$, $CH_4(g) - H_2O(g)$, $H_2O(g) - H_2(g)$ equilibria, as well as $1000 \times \ln {}^{13}\alpha^{eq}_{CH_4(g)-CO_2(g)}$ — all based on the DBO-corrected PIMC calculations were fit as a function of $1/T$ using a $4^{th}$ order polynomial (Fig. 5.2, see Table 5.5 for the coefficients). Although ANOVA indicates higher order terms are statistically significant ($p < 0.05$) in some cases, their inclusion changed values of $1000 \times \ln \alpha$ by less than 0.5‰ for hydrogen isotopes and less than 0.05‰ for carbon isotopes in all cases across the calculated range in temperatures. We consider this insignificant and limit the fits to $4^{th}$ order terms for simplicity.

| T (°C) | CH$_4$/H$_2$ | CH$_4$/H$_2$O | H$_2$O/H$_2$ | T (°C) | CH$_4$/CO$_2$ |
|---|---|---|---|---|---|
| 0.98 | 0.3 | 33.9 | -33.5 | 1.85 | -0.26 |
| 2.94 | 0.4 | 33.6 | -33.3 | 11.85 | -0.25 |
| 9.9 | 0.3 | 32.7 | -32.4 | 26.85 | -0.24 |
| 14.57 | 0.3 | 32.3 | -31.9 | 51.85 | -0.23 |
| 25.4 | 0.3 | 31.1 | -30.8 | 76.85 | -0.21 |
| 35.4 | 0.3 | 30 | -29.8 | 101.85 | -0.2 |
| 50.9 | 0.3 | 28.7 | -28.3 | 126.85 | -0.18 |
| 75.65 | 0.2 | 26.6 | -26.3 | 176.85 | -0.16 |
| 99.9 | 0.3 | 24.9 | -24.6 | 226.85 | -0.14 |
| 127.4 | 0.2 | 23.2 | -23 | 276.85 | -0.13 |
| 157.35 | 0.2 | 21.5 | -21.4 | 326.85 | -0.12 |
| 180 | 0.2 | 20.5 | -20.3 | 376.85 | -0.11 |
| 200.2 | 0.2 | 19.6 | -19.4 | 426.85 | -0.1 |
| 203.35 | 0.2 | 19.5 | -19.3 | 476.85 | -0.1 |
| 219.88 | 0.2 | 18.8 | -18.7 | 526.85 | -0.09 |
| 264.07 | 0.2 | 17.3 | -17.1 | 601.85 | -0.08 |
| 302.55 | 0.2 | 16.1 | -16 | 801.85 | -0.07 |
| 364.58 | 0.2 | 14.6 | -14.4 | 1001.85 | -0.06 |
| 407 | 0.1 | 13.6 | -13.5 | 1301.85 | -0.05 |
| 458.1 | 0.1 | 12.7 | -12.6 | | |
| 476.5 | 0.1 | 12.4 | -12.3 | | |
| 503.4 | 0.1 | 11.9 | -11.9 | | |

Table 5.4: Effect of the DBO correction on $1000 \times \ln{}^D\alpha^{eq}$ and $1000 \times \ln{}^{13}\alpha^{eq}$ values. The value without the DBO correction is subtracted from the value with the DBO correction and the result is shown in the table.

| $1000 \times \ln \alpha^{eq}$ | $1/T^4$ (K$^{-4}$) | $1/T^3$ (K$^{-3}$) | $1/T^2$ (K$^{-2}$) | $1/T$ (K$^{-1}$) | const |
|---|---|---|---|---|---|
| CH$_4$ – H$_2$ | 3.6652E+12 | -4.5802E+10 | 2.2230E+08 | -1.4941E+04 | -41.20 |
| H$_2$O – H$_2$ | 2.2218E+12 | -2.6028E+10 | 1.1628E+08 | 2.2826E+05 | -115.00 |
| CH$_4$ – H$_2$O | 1.4611E+12 | -1.9936E+10 | 1.0656E+08 | -2.4396E+05 | 74.20 |
| CH$_4$ – CO$_2$ | -2.6547E+11 | 2.8888E+09 | -1.3306E+07 | 1.8968E+03 | -0.14 |

Table 5.5: Coefficients of 4$^{th}$ order polynomial fits to DBO-corrected PIMC calculations of $1000 \times \ln \alpha^{eq}$ for the pairs of gases.


**Determination of $^D\alpha^{eq}_{CH_4(g)-H_2O(l)}$ vs. temperature at isotopic equilibrium**

In order to calculate the equilibrium fractionation factor between CH$_4(g)$ and liquid water ($^D\alpha^{eq}_{CH_4(g)-H_2O(l)}$) using our experimental calibration for $^D\alpha_{CH_4(g)-H_2(g)}$, it is necessary to know $^D\alpha_{H_2O(l)-H_2(g)}$ as a function of temperature for an isotopically equilibrated system. The hydrogen-isotope fractionation factor between gaseous H$_2$ and liquid water has been determined experimentally from 6 to 95°C.[181] Additionally, equilibrium hydrogen-isotope fractionation factors between gaseous H$_2$ and water vapor ($^D\alpha_{H_2O(g)-H_2(g)}$) have been determined experimentally from 80

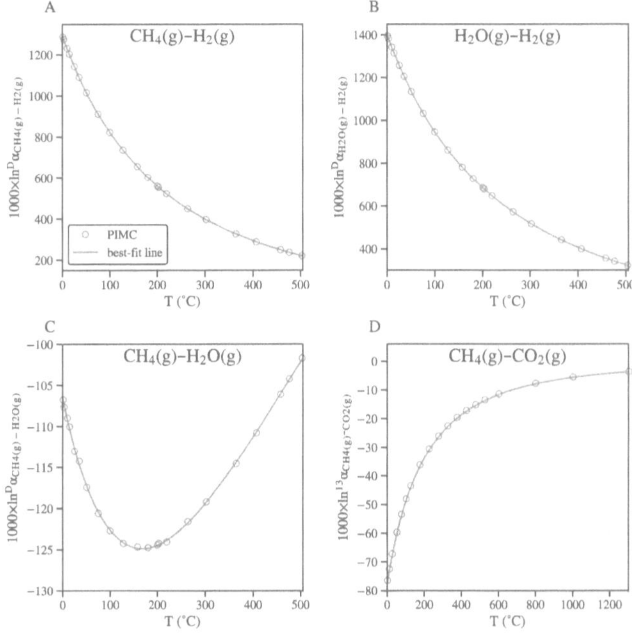

Figure A1: Results from DBO-corrected PIMC calculations for (A) $1000\times\ln^D\alpha_{CH4(g)-H2(g)}$, (B) $1000\times\ln^D\alpha_{H2O(g)-H2(g)}$, (C) $1000\times\ln^D\alpha_{CH4(g)-H2O(g)}$, and (D) $1000\times\ln^{13}\alpha_{CH4(g)-CO2(g)}$ vs. temperature (°C). Lines are 4th order polynomial fits to the calculated data points vs. $1/T$ (K⁻¹). Calculated values are given in Tables 3 and 4. The terms for each best-fit line and the associated error is given in Table 5.5.

Figure 5.2: **DBO-corrected PIMC calculations** of $1000\times\ln^D\alpha$ (A-C) and $1000\times\ln^{13}\alpha$ (D) vs. temperature (°C). Lines are 4th order polynomial fits to the calculated data points vs. $1/T$ (K$^{-1}$). Calculated values are given in Tables 5.2 and 5.3. The terms for each best-fit line are given in Table 5.5.

to 200°C[182] and from 51 to 742°C.[183] We use these experiments to construct a calibration of $^D\alpha_{H_2O(l)-H_2(g)}$ as a function of temperature. To do this, we convert experimentally determined values of $^D\alpha_{H_2O(g)-H_2(g)}$ to $^D\alpha_{H_2O(l)-H_2(g)}$ using the experimentally based calibration for hydrogen isotopic equilibrium between water vapor and liquid water ($^D\alpha_{H_2O(l)-H_2O(g)}$).[167]

In order to calculate a fit to the experimental data vs. temperature, we first average published experimental values of $1000\times\ln^D\alpha_{H_2O(g)-H_2(g)}$ or $1000\times\ln^D\alpha_{H_2O(l)-H_2(g)}$ performed within 1°C of each other from a given study. We take these averages to prevent experiments replicated at the same temperature from having undue weight in our fits. Following this, we convert $1000\times\ln^D\alpha_{H_2O(g)-H_2(g)}$ to $1000\times\ln^D\alpha_{H_2O(l)-H_2(g)}$ using the polynomial fit (the first equation in the abstract of[167]). Note that we only used data from for experiments performed below the critical point of water (374°C). The compiled experimental data is displayed in Fig. 5.3A. Generally, when experiments from different studies were performed over a similar temperature range ($\sim$50 to 200°C), there is agreement between all three studies. As previously noted[181,184], two data points from [183] at 59 and 64°C differ both from other data in that calibration as well as those from [181] over the same temperature range. These points are noted in the figure and are omitted from our calibration due to the apparent anomalous behavior.[181]

We initially calculated the temperature dependence of $1000 \times \ln {}^{D}\alpha_{H_2O(l)-H_2(g)}$ vs. $1/T$ (K$^{-1}$) using a 4$^{th}$ order polynomial fit to the experimental data given in Figure 5.3A. However, the best-fit line had multiple changes in concavity. In contrast, theory predicts a smooth, concave up temperature dependence. We believe this difference is due to the polynomial fit being influenced by the scatter of the experimental data, which is approximately ±25‰ from 0 to 374°C.

In order to use the experimental data as a constraint on the temperature dependence of $1000 \times \ln {}^{D}\alpha_{H_2O(l)-H_2(g)}$, but at the same time avoid fitting the experimental noise, we do not directly regress these experimentally determined equilibrium $1000 \times \ln \alpha$ values vs. $1/T$; instead, we perform a least squares fit of the theoretical DBO-corrected PIMC calculations to the experimental data, where only a constant (temperature-independent) term is allowed to vary. This is done to ensure the theoretically expected shape of the temperature dependence for $1000 \times \ln {}^{D}\alpha_{H_2O(l)-H_2(g)}$ is preserved in the fit to the experimental data. To do this we combined our DBO-corrected PIMC calculations of $1000 \times \ln {}^{D}\alpha^{eq}_{H_2O(g)-H_2(g)}$ at the experimental temperatures with those of $1000 \times \ln {}^{D}\alpha_{H_2O(l)-H_2(g)}$ using the calibration of [167]. A constant offset is then added to the theoretical values based on the least square fit to the experiment. We then found the final equation for $1000 \times \ln {}^{D}\alpha_{H_2O(l)-H_2(g)}$ as a function of temperature as follows: we calculated $1000 \times \ln {}^{D}\alpha^{eq}_{H_2O(g)-H_2(g)}$ based on our 4$^{th}$ order fits to DBO-corrected PIMC calculations from 0 to 374°C at 0.1°C intervals, added the offset term found above, and then added the $1000 \times \ln {}^{D}\alpha_{H_2O(l)-H_2O(g)}$ value calculated at that temperature based on the calibration of [167]. We then fit a 4$^{th}$ order polynomial to these points. This yielded the following equation, which is explicitly valid from 6.7 to 357°C (i.e., the experimental temperature range):

$$1000 \times \ln {}^{D}\alpha_{H_2O(l)-H_2(g)} = \frac{7.9443 \times 10^{12}}{T^4} - \frac{8.7772 \times 10^{10}}{T^3} + \frac{3.8504 \times 10^{8}}{T^2} - \frac{2.6650 \times 10^{5}}{T} + 202.57.$$

(5.2)

The PIMC calculations were offset by 0.49‰ ($\pm$ 1.91‰, 1 s.e.) to fit the experimental data (Fig. 5.3C). Although the offset required for the DBO-corrected PIMC fit is within 1 s.e. of 0, we still apply this offset (0.49‰) in Eq. (5.2) for consistency. For this fit, the standard deviation of the residual from all experiments is 12.3‰. The mean and ±1$\sigma$ of residuals vs. Eq. (5.2) for each individual published data set are: 1.3 ± 5.2‰ for [182], -2.6 ± 15.4‰ for [183] and 2.3 ± 10.1‰ for [181] (Fig. 5.3B). Finally, we note that inclusion of the two omitted data points from [183] discussed above changes the constant term added to our PIMC calculations by -2.7‰ (-2.16‰ vs. 0.49‰).

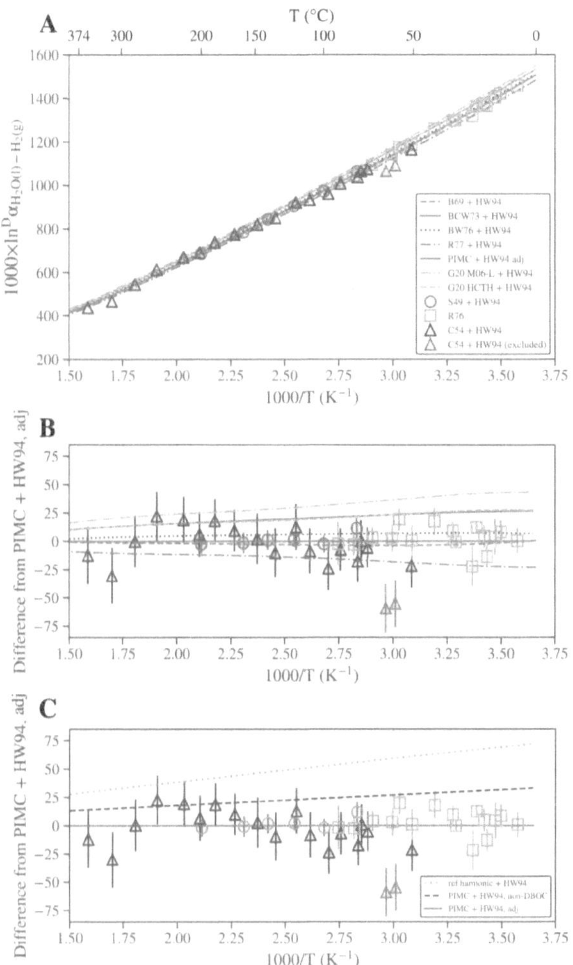

Figure 4: (A) Experimental and theoretical determinations of equilibrium $1000\times\ln^{D}\alpha_{H2O(l)-H2(g)}$ vs. 1000/T ($K^{-1}$). Theoretical estimates are calculations from Bottinga (1969) (B69); Bron et al. (1973) (BCW73); Bardo and Wolfsberg (1976) (BW76); Richet et al. (1977) (R77); and Gropp et al. (in press) (G20 M06-L; G20 HCTH). Experimental data are given as averages, as discussed in Section 3.4. Experimental data from Suess (1949) (S49) and Cerrai et al. (1954) (C54) were measured as $1000\times\ln^{D}\alpha_{H2O(g)-H2(g)}$ and transformed to $1000\times\ln^{D}\alpha_{H2O(l)-H2(g)}$ using the experimental calibration of $1000\times\ln^{D}\alpha_{H2O(l)-H2O(g)}$ vs. temperature from Horita and Wesolowski (1994) (HW94). Data from Rolston et al. (1976) (R76) are direct experimental determinations of $1000\times\ln^{D}\alpha_{H2O(l)-H2(g)}$. Two data points from Cerrai at al. (1954) are excluded from our fits to the experimental data (red triangles; see discussion in Section 3.5). (B) Differences between theoretical calculations and experimental data vs. our PIMC adjusted ('adj') best-fit line to the experimental data (eq. 8; PIMC + HW94 adjusted). Same legend as (A). This best-fit line was

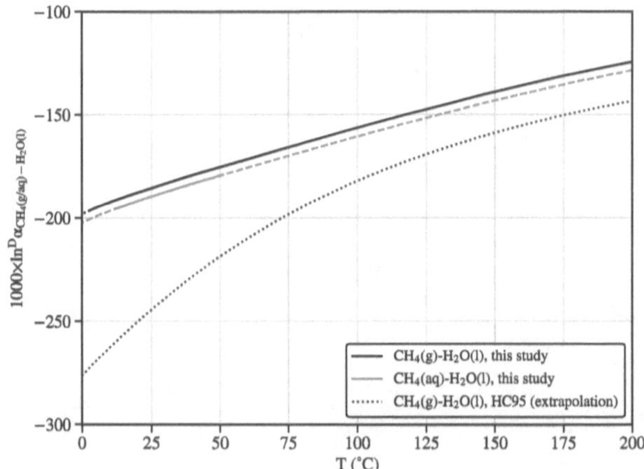



2346
2347 Figure 5: 1000×ln$^D\alpha_{CH4(g/aq)-H2O(l)}$ vs. temperature from this study and from Horibe and Craig (1995) [HC95
2348 (extrapolation)]. Dotted and dashed lines indicate extrapolations beyond experimentally calibrated

**Figure 5.4: Comparison of** $1000 \times \ln{}^D\alpha_{CH_4(g/aq)-H_2O(l)}$ **vs. temperature from this study and from [140]**, labelled as HC95 (extrapolation). Dotted and dashed lines indicate extrapolations beyond experimentally calibrated temperatures.

The residuals of fits show no obvious pattern either as a function of temperature or study, scattering about a value of 0‰ within error for all studies (Fig. 5.3B and C). This indicates that the experiments and theory agree in terms of the expected temperature dependence and justifies our approach to using the theoretical temperature dependence as a basis for our fit to the experiments. Additionally, it shows that experiments from different studies are in agreement when they overlap in temperature. We note that this is in contrast with the conclusions of [140] and [119]. They considered these experimental calibrations to be in disagreement at high temperatures (>300°C)[140] and low temperatures (<100°C).[119] In Figure 5.3A, B, and C, we also include other calculation and experimental values of $1000 \times \ln{}^D\alpha_{H_2O(l)-H_2(g)}$.[90,144,166,184,185]

**Determination of** ${}^D\alpha_{CH_4(g)-H_2O(l)}$ **and of** ${}^D\alpha_{CH_4(aq)-H_2O(l)}$ **as a function of temperature at isotopic equilibrium**

We derived an equation for $1000 \times \ln{}^D\alpha_{CH_4(g)-H_2O(l)}$ vs. $1/T$ (K$^{-1}$) at isotopic equilibrium by subtracting our Eq.( 5.2) from our experimental equation for $1000 \times \ln{}^D\alpha_{CH_4(g)-H_2(g)}$ (Eq. 5.3):

$$1000 \times \ln{}^D\alpha_{CH_4(g)-H_2(g)} = \frac{3.5317 \times 10^7}{T^2} + \frac{2.7749 \times 10^5}{T} - 179.48. \qquad (5.3)$$

This yields Equation (5.4) that can be interpolated from 3 to 200°C (Fig. 5.4).

$$1000 \times \ln{}^{D}\alpha_{CH_4(g)-H_2O(l)} = -\frac{7.9443\times10^{12}}{T^4} + \frac{8.7772\times10^{10}}{T^3} - \frac{3.4972\times10^8}{T^2} + \frac{5.4399\times10^5}{T} - 382.05.$$

(5.4)

We use 3°C as our lower limit as this is the lowest temperature at which the hydrogen isotopes of $CH_4$ and $H_2$ were equilibrated experimentally. We note that the lowest temperature available for $H_2O(l) - H_2(g)$ hydrogen isotope equilibrium is 6.7°C. However, as our calibration of $1000 \times \ln{}^{D}\alpha_{H_2O(l)-H_2(g)}$ is based on a theoretical fit to the experimental data, we consider it acceptable to use Eq. (5.4) to calculate equilibrium $1000 \times \ln{}^{D}\alpha_{CH_4(g)-H_2O(l)}$ values from 3 to 200°C.

Finally, $CH_4$ is found in the environment both as a gas and dissolved in water. An experimental determination of equilibrium ${}^{D}\alpha_{CH_4(aq)-CH_4(g)}$ values can be derived from experimental determinations of $1000 \times \ln(CD_4/CH_4)$ in gas vs. liquid given in [186] from 12-51°C. If we assume a random hydrogen isotopic distribution, the $4^{th}$ root of gaseous or dissolved $CD_4/CH_4$ ratios allows for the calculation of equilibrium $1000 \times \ln{}^{D}\alpha_{CH_4(aq)-CH_4(g)}$ values. These solubility isotope effects show no apparent temperature dependence from 12-51°C —- linear regression of $1000 \times \ln{}^{D}\alpha_{CH_4(aq)-CH_4(g)}$ vs. $1000/T$ $(K^{-1})$ yields a slope of -1.3 ± 2.6 (1 s.e.). Based on this, we assume the value of $1000 \times \ln{}^{D}\alpha_{CH_4(aq)-CH_4(g)}$ from 12 to 51°C is constant and equal to 4.2 ± 0.4‰ (1 s.e.). This is the average of the experimental determinations of $1000 \times \ln{}^{D}\alpha_{CH_4(aq)-CH_4(g)}$ at all temperatures.

Incorporating this isotope effect into Eq. (5.4) yields the following equation for $1000 \times \ln{}^{D}\alpha_{CH_4(aq)-H_2O(l)}$ vs. $1/T$ $(K^{-1})$ at isotopic equilibrium:

$$1000 \times \ln{}^{D}\alpha_{CH_4(aq)-H_2O(l)} = -\frac{7.9443\times10^{12}}{T^4} + \frac{8.7772\times10^{10}}{T^3} - \frac{3.4972\times10^8}{T^2} + \frac{5.4399\times10^5}{T} - 386.25.$$

(5.5)

This equation can be interpolated from 12-51°C (Fig. 5.4).

**Determination of ${}^{13}\alpha_{CH_4(g)-CO_2(g)}$ and ${}^{13}\alpha_{CH_4(aq)-CO_2(aq)}$ at isotopic equilibrium as a function of temperature**

Here we present equations for carbon isotopic equilibrium between $CH_4(g)$ and $CO_2(g)$ and between $CH_4(aq)$ and $CO_2(aq)$ vs. temperature. We are aware of two studies experimentally determined the equilibrium fractionation factors between $CH_4(g)$ and $CO_2(g)$ $({}^{13}\alpha_{CH_4(g)-CO_2(g)})$.[162,163] Between these two studies, experimentally determined values of ${}^{13}\alpha_{CH_4(g)-CO_2(g)}$ for isotopic equilibrium are available from 200 to 1200°C and are in agreement (within 0.01 to 1.01‰) where experiments overlap in temperature (300 to 600°C).

We are interested here in values for ${}^{13}\alpha_{CH_4(g)-CO_2(g)}$ at isotopic equilibrium at temperatures <200°C, i.e., overlapping the temperature range of the $CH_4 - H_2$ equilibration experiments performed in this study. As no experimental calibrations are available at these temperatures, we estimate these values using the approach

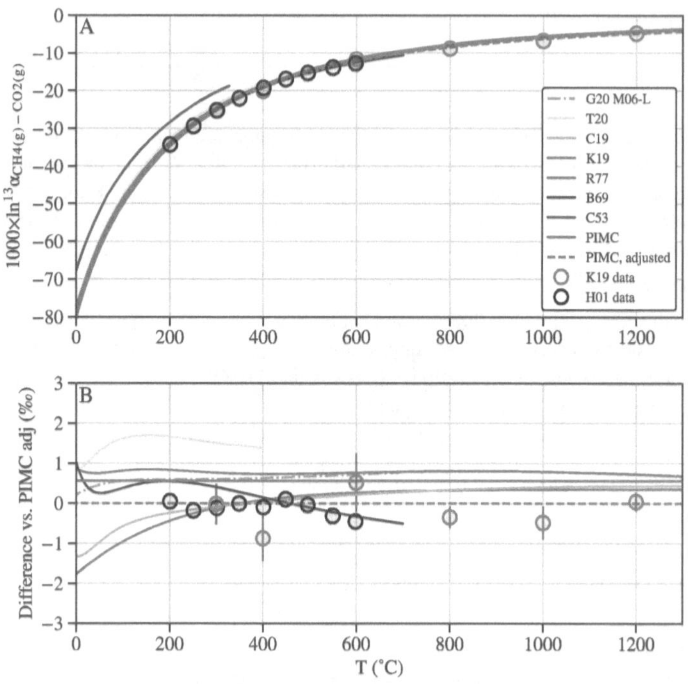



Figure 5.5: $1000 \times \ln^{13}\alpha_{CH_4(g)-CO_2(g)}$ **from published experimental fits**: [163] (K19), theory [159] (C53), [144] (B69), [90] (R77), [165] (C19), [155] (T20) and [166] (G20 M06-L), new calculation done in this study (PIMC, adjusted), and experimental data used to adjust the theoretical fits, as discussed.[162,163] (B) Difference between $1000 \times \ln^{13}\alpha_{CH_4(g)-CO_2(g)}$ from the various studies vs. our adjusted PIMC line. [159] is off scale. Error is 1 s.e.; when not shown, error is smaller than symbol.

Figure A.2: $1000\times\ln^{13}\alpha_{CH_4-CO_2}$ vs. temperature (°C) showing estimates for $1000\times\ln^{13}\alpha_{CH_4-CO_2(g)}$ from published experimental fits (Kueter et al., 2019 (K19)), theory (Craig, 1953 (C53); Bottinga, 1969 (B69); Richet et al., 1977 (R77); Chen et al., 2019 (C19); Thiagarajan et al., 2020 (T20); Gropp et al., in press (G20 M06-L)), new theoretical calculations done in this study (PIMC (PIMC, adjusted)), and experimental data used to adjust the theoretical fits, as discussed in Section 3.8 (Horita, 2001; Kueter et al., 2019). (B): Difference between $1000\times\ln^{13}\alpha_{CH4(g)-CO2(g)}$ from the various studies vs. our adjusted PIMC line. Craig (1953) is off scale. Error is 1 s.e.; when not shown, error is smaller than symbol.

discussed above in which a theoretical calibration is offset to fit the experimental data and used as the basis for extrapolations to lower temperatures.[164] Specifically, we use the fourth order polynomial fit of $1000 \times \ln^{13}\alpha_{CH_4(g)-CO_2(g)}$ vs. $1/T$ (K$^{-1}$) to our theoretically calculated points, and then fit these curves to the experimental data using a constant offset to minimize the sum of square residuals as above. This results in the following equation (Fig. 5.5):

$$1000 \times \ln^{13}\alpha_{CH_4(g)-CO_2(g)} = -\frac{2.6636\times10^{11}}{T^4} + \frac{2.8883\times10^9}{T^3} - \frac{1.3292\times10^7}{T^2} + \frac{1.7783\times10^3}{T} - 0.70.$$

$$(5.6)$$

78

For our DBO-corrected PIMC calculations, the theoretical curve is decreased by

0.56‰ (±0.12‰, 1 s.e.) to fit the experimental data. Uncorrected PIMC calculations yield an offset of 0.94‰ (±0.08‰, 1 s.e.).

We account for isotope effects associated with the dissolution of $CO_2$ and methane in water using the experiments from [187] in which $CO_2(g)$ and $CO_2(aq)$ were equilibrated from 0 to 60°C. We refit the data in terms of $1000 \times \ln^{13} \alpha_{CO_2(aq)-CO_2(g)}$ vs. $1/T$ ($K^{-1}$) (the fit was originally given as $1000 \times \left(^{13}\alpha_{CO_2(aq)-CO_2(g)} - 1\right)$ vs. T(°C)) to allow for the interconversion of $\ln \alpha$ calibrations of various species via addition (or subtraction) of polynomial terms. This refit results in the following equation:

$$1000 \times \ln {}^{13}\alpha_{CO_2(aq)-CO_2(g)} = -\frac{378.46}{T} + 0.2016. \tag{5.7}$$

For methane, [188] provides values for experimental determinations of $1000 \times \ln {}^{13}\alpha_{CH_4(aq)-CH_4(g)}$ at isotopic equilibrium at 20, 50, and 80°C that were originally presented in [189]. We used this data to find the following equation:

$$1000 \times \ln {}^{13}\alpha_{CH_4(aq)-CH_4(g)} = \frac{485.54}{T} - 1.0453 \tag{5.8}$$

Combination of Eqs. (5.6), (5.7) and (5.8) yields the following equation for carbon isotopic equilibrium between aqueous $CH_4$ and $CO_2$:

$$1000 \times \ln {}^{13}\alpha_{CH_4(aq)-CO_2(aq)} = -\frac{2.6636 \times 10^{11}}{T^4} + \frac{2.8883 \times 10^9}{T^3} - \frac{1.3292 \times 10^7}{T^2} + \frac{2.6423 \times 10^3}{T} - 1.95 \tag{5.9}$$

This resulting equation is interpolatable from 20 to 60°C.

## 5.4 Discussion

In Figure 5.4, we compare our experimentally based calibration of $^D\alpha_{CH_4(g)-H_2O(l)}$ for isotopic equilibrium to the experimentally based calibration of [140] for temperatures from 0 to 200°C. Over this temperature range, the two calibrations show disagreement, the magnitude of which increases with decreasing temperature. For example, at 3°C (our lowest temperature experiment), the two calibrations disagree by 76‰. Our calibration indicates that methane formed in the near-surface (for example, 0 to 50°C) should be between ~200‰ and ~175‰ lower in $\delta D$ compared to the source waters vs. ~275‰ to ~220‰ based on the calibration of [140].

**Comparison of our experimentally determined** $1000 \times \ln {}^D\alpha_{CH_4(g)-H_2(g)}$ **fractionation factor to calculation**

In Figure 5.6C, we compare the differences between $1000 \times \ln {}^D\alpha_{CH_4(g)-H_2(g)}$ predicted based on the best-fit line to our experimental data (Eq. 5.3) v s. those from previous theoretical studies, our PIMC calculations, and the extrapolation of the calibration given by [140]. Visually, our theoretical PIMC calculations for isotopic

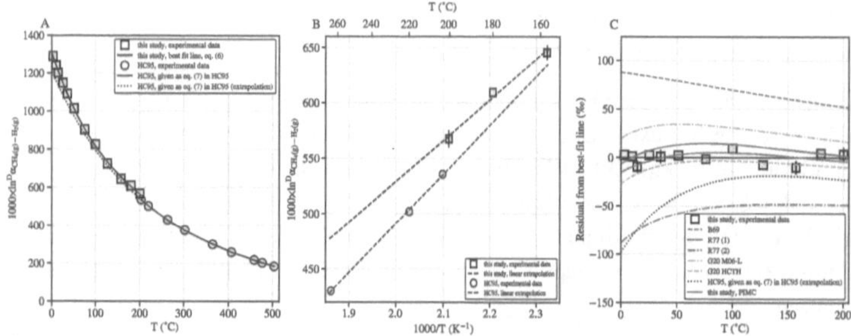

Figure 6: (A) Comparisons of determinations of 1000×ln$^D\alpha_{CH_4(g)-H_2(g)}$ from this study vs. from Horibe and Craig (1995) (HC95). (B) Experimental determinations vs. temperature (°C) from this study and Ref. [140] (HC95) ... linear extrapolations vs. 1/T (K$^{-1}$) through the given points and are provided as guidance. The 22.9‰ offset between the two studies at 200°C is also observed at lower and higher temperatures as seen by the extrapolations. (C) Relative differences of 1000×ln$^D\alpha_{CH_4(g)-H_2(g)}$ between various theoretical calculations and HC95 ... of HC95 vs. the best-fit line to our data ... the equation given in Horibe and Craig (1995; eq. 7 of that study) to temperatures below 200°C; B69 is Bottinga (1969); R77 is Richet et al. (1977), ... harmonic approximation ... anharmonic approximation. G20 M06-L is M06-L calculated RPFR as presented in Gropp et al. (in press), while G20 HCTH is HCTH calculated ... this study, PIMC ... theoretical studies, the presented lines are 4th order polynomial fits of 1000×ln$^D\alpha_{CH_4(g)-H_2(g)}$ vs. 1/T based on calculated values of RPFRs from the given study. Error bars for experimental points shown are ±1 s.e. and are smaller than the symbol when not shown.

Figure 5.6: **Comparisons of** $1000 \times \ln^D\alpha_{CH_4(g)-H_2(g)}$ (A) from this study and Ref. [140] (HC95). (B) Experimental determinations vs. temperature (°C) from this study and Ref. [140] (HC95) from 150 to 275°C. Lines are linear extrapolations vs. $1/T$ (K$^{-1}$) through the given points and are provided as guidance. The 22.9‰ offset between the two studies at 200°C is also observed at lower and higher temperatures as seen by the extrapolations. (C) Relative differences between various calculation and the extrapolation of HC95 vs. the best-fit line to our data (Eq. 5.3). HC95 (extrapolation) is the extrapolation of the equation given in Ref. [140], (Eq. 7 of that study) to temperatures below 200°C; B69 is from [144], R77 is from [90], where (1) represents and estimate where the excess factor $X_{CH_4}$ is calculated using a harmonic approximation while R77 (2) indicates an anharmonic approximation. G20 M06-L is M06-L and HCTH are from Ref. [166]. "this study, PIMC" is the polynomial fit to the DBO-corrected PIMC calculation presented here. For the theoretical studies, the presented lines are 4$^{th}$ order polynomial fits of $1000 \times \ln^D\alpha_{CH_4(g)-H_2(g)}$ vs. $1/T$ based on calculated values of RPFRs from the given study. Error bars for experimental points shown are ±1 s.e. and are smaller than the symbol when not shown.

equilibrium best match our experimental data. For example, the largest disagreement between the fit to the PIMC calculations and the fit to the experimental data is 7‰, at 3°C (disagreement between individually calculated and measured points is up to 16.5‰). We show this agreement quantitatively with a plot of $1000 \times \ln^D\alpha^{eq}_{CH_4(g)-H_2(g)}$ of PIMC calculations vs. our experiments performed at same temperature (Fig. 5.7). The slope of the best-fit line is 0.986 ± 0.009 (1 s.e.) with an intercept of 10.99 ± 8.81 (1 s.e.). Thus, these determinations of $1000 \times \ln^D\alpha_{CH_4(g)-H_2(g)}$ agree within 2 s.e. of the 1:1 line both in slope and intercept. Put another way, satisfyingly, our experimental results agree 1:1 with our calculation at the highest levels of theory (PIMC with DBO corrections) as yet employed for these calculations.

Other calculation of $^D\alpha_{CH_4(g)-H_2(g)}$ also visually agree with our experimental data, although not as closely as the calculations presented in this work. These include one set of calculations from [90] in which $X_{CH_4}$, their so-called "excess factor," is calculated using a harmonic approximation (labeled as R77 (1) in Fig. 5.6C), which shows a maximum difference of 18‰ vs. the fit to our experimental data.

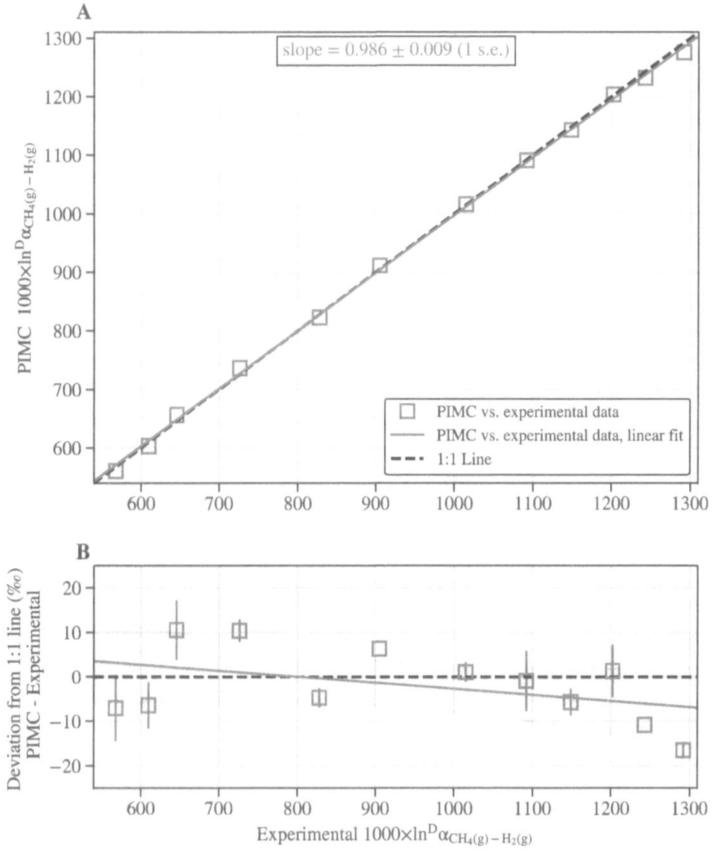


Figure 5.7: **Comparison of DBO-corrected PIMC calculation of** $1000 \times \ln {}^D\alpha_{CH_4(g)-H_2(g)}$ **vs. our experimental results.** The best-fit line yields a slope of 0.986 ± 0.009 (1 s.e.) and intercept of 10.99 ± 8.81 (1 s.e.). Thus, the two are in 1:1 agreement within 2 s.e. (B) Deviation of calculation vs. experimental determinations of $1000 \times \ln {}^D\alpha_{CH_4(g)-H_2(g)}$ relative to a 1:1 line. Shading indicates 95% confidence interval for the linear fit. Y-axis error bars are 1 s.e. error for both PIMC and our experimental data propagated in quadrature. 1 s.e. errors smaller than the size of the symbol are omitted (which includes all error in the X-axis in addition to several Y-axis error bars)

Figure 7. (A) Comparison of DBO-corrected PIMC theoretical calculations of 1000×ln$^D\alpha_{CH4(g)-H2(g)}$ vs. our experimental results. The best-fit line yields a slope of 0.986 ± 0.009 (1 s.e.) and intercept of 10.99 ± 8.81 (1 s.e.). Thus, the two are in 1:1 agreement within 2 s.e. (B) Deviations of theoretical calculations vs. experimental determinations of 1000×ln$^D\alpha_{CH4(g)-H2(g)}$ relative to 1:1 line. Shading indicates 95% confidence interval for the linear fit. Y-axis error bars are 1 s.e. error for both PIMC and our experimental data propagated in quadrature. 1 s.e. errors smaller than the size of the symbol are omitted (which includes all error in the X-axis in addition to several Y-axis error bars)

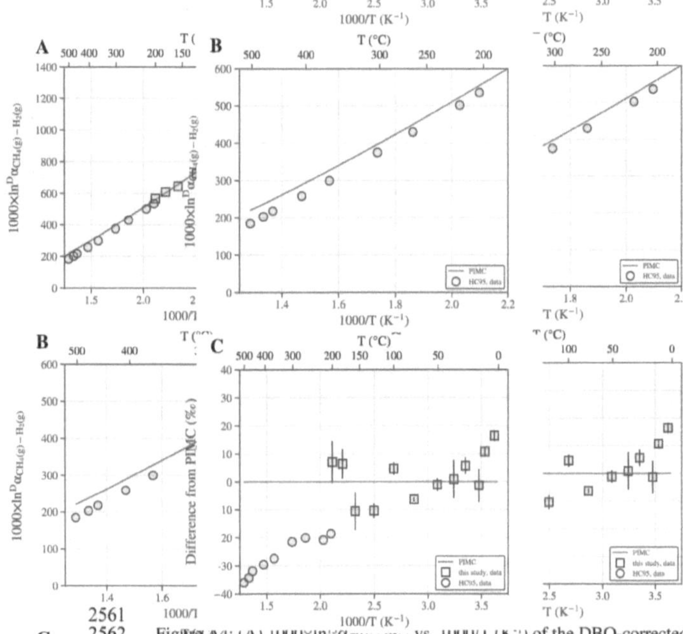

Figure 5.8: **Comparison of** $1000 \times \ln ^D\alpha_{CH_4(g)-H_2(g)}$ **from this study to the experimental data from [140] (HC95).** (B) Zoom-in showing just HC95 data and the PIMC fit from this study. Note the offset between HC95 data and the theoretical line. (C) Difference plot of data from this study and from HC95 relative to PIMC calculations from this study. The PIMC calculations were done at the same experimental temperatures for both our experimental results and those from Horibe and Craig (1995). Error bars are 1 s.e..

Additionally, recent calculations differ from the fit to our experimental data by at most +34‰ (referred to by them as M06-L) or -26‰ (referred to by them as HCTH).[166] The other theoretical calibrations show larger differences: the curve for the calculations from [144] is up to 95‰ greater than the fit to our experimental data, while that from [90] in which $X_{CH_4}$ is calculated with anharmonic corrections (labeled as R77(2) in Fig. 5.6C) is up to 87‰ lower at low temperatures for both our experimental results and those from Horibe and Craig (1995). Error bars are 1 s.e..

Finally, in Figure 5.8 we compare $1000 \times \ln ^D\alpha_{CH_4(g)-H_2(g)}$ from [140] vs. our DBO-corrected PIMC results. Data from [140] are offset to lower values than those based on theory from 18 to 37‰. In addition to the offset itself, the degree of offset is a function of temperature (increasing with increasing temperature) indicating that there is a difference in the temperature dependence between PIMC calculations from this study and the data from [140].

**Comparison of experimental vs. theoretical** $1000 \times \ln {}^{D}\alpha_{\mathrm{H_2O}(l)-\mathrm{H_2}(g)}$ **for isotopic equilibrium**

Calculations of equilibrium $1000 \times \ln {}^{D}\alpha_{\mathrm{H_2O}(l)-\mathrm{H_2}(g)}$ values are compared to experimental data as a function of temperature in Fig. 5.3. We note that in all cases, calculations are for $1000 \times \ln {}^{D}\alpha_{\mathrm{H_2O}(g)-\mathrm{H_2}(g)}$ and are converted to $1000 \times \ln {}^{D}\alpha_{\mathrm{H_2O}(l)-\mathrm{H_2}(g)}$ using the experimental calibration of equilibrium $\alpha_{\mathrm{H_2O}(l)-\mathrm{H_2O}(g)}$ values.[167] Our calculations of equilibrium $1000 \times \ln {}^{D}\alpha^{eq}_{\mathrm{H_2O}(l)-\mathrm{H_2}(g)}$ values based on the PIMC methodology with the DBO correction are in close agreement with the experimental data. Using the minimization scheme described above (Fig. 5.3), the DBO-corrected PIMC calculations are offset by 0.49‰ from the experimental data. In contrast, we observe a +24.5‰ offset between PIMC calculations without the DBO correction and experimental data. Additionally, the harmonic RPFR calculated here using the same reference potentials ("reference harmonic" line) as used for PIMC calculations is offset by +54.4‰ from the experimental data. Offsets between experiments and other calculations which include anharmonic corrections to harmonic RPFRs are as follows: +2.7‰ from [144]; +21.1‰ for [185], -16.3‰ for [90], and +6.1‰ from [184]. Calculations presented in [166], which do not include anharmonic corrections, are offset by +32.9‰ (M06-L) and +21.0‰ (HCTH) from the experimental data. We note that the calculations presented in [184] also include an adiabatic correction to the Born-Oppenheimer (B-O) approximation,[105,190] which is just a different name for the same correction as the DBO correction presented in this work. Here, we base our correction off of values calculated by[174], which are based on more rigorous levels of theory and larger basis sets.

The observed differences between the various calculations vs. experimentally measured values of equilibrium fractionation factors may be due to inaccuracies in calculations or aspects of the experiments. We note that as the calculations do not all agree for a given set of molecules, they cannot all be accurate. Here we examine the sensitivity of calculated RPFRs to a range of factors, including: (i) level of electronic structure theory; (ii) vibrational anharmonicity; and (iii) DBO corrections.

**Accuracy of the electronic structure method**

Accurate calculations of RPFRs necessarily require the use of accurate potential energy surfaces. However, obtaining accurate potential energy surfaces remains challenging even for the small molecules we are investigating. For the PIMC calculations we used high-quality reference potentials as described in Section 5.3. However, these potentials are not exact, and small inaccuracies in electronic energy could result in inaccurate RPFRs. There are two issues that need to be addressed to validate the accuracy of these potentials: electronic correlation effects and basis set completeness (i.e., achieving the complete basis set or CBS limit).

We explored the magnitude of the error resulting from the inaccuracies in reference potentials at equilibrium geometry as follows. Harmonic RPFRs were computed using Eq. (3.25) based on the normal mode frequencies obtained with either increasingly refined treatment of electronic correlation using the same basis set or using the same electron-correlation treatment with increasing basis set size. These calculations were then compared to the harmonic result, obtained using the reference potentials. We interpret convergence of harmonic RPFRs with respect to the level of theory or basis set size to indicate sufficient accuracy of the electronic structure methods. Finally, the level of agreement between the converged result and that obtained from the reference harmonic potentials is used as an estimate of the possible size of inaccuracies that result from the reference potentials used in this study (Section 5.3).

We computed the harmonic frequencies using the RHF, MP2, CCSD, CCSD(T) levels of theory (abbreviations defined in Section 5.3; $H_2$ cannot be calculated at CCSD(T) level, as it is exact as CCSD). For these calculations, we used the same high-quality basis set (aug-cc-pVQZ) throughout. For the basis set convergence test, harmonic frequencies were computed at the CCSD(T) level of theory using the following standard and augmented correlation-consistent polarized valence basis sets: cc-pVXZ and aug-cc-pVXZ. Here X denotes the progression of the basis set size: D for double-, T for triple- and Q for quadruple-$\zeta$.[171,175] The electron-electron correlation test results are given in Figure 5.9 and the basis set size test results in Figure 5.10. We plot $\delta$ RPFR on the y-axis, where $\delta$ RPFR $= 1000 \times \ln\left(RPFR_X/RPFR_{\text{reference}}\right)$ and is reported in ‰. X denotes the value of the variable tested and the reference is the RPFR calculated using harmonic frequencies from the same PES as used in the PIMC calculations (the so-called reference harmonic lines; Section 5.3). This choice of reference is not meant to indicate that the reference results are necessarily the most accurate but is used simply for the purpose of comparison. The RHF results in Figure 5.10 are scaled down by a factor of 10 in order to fit on the same axis. Focusing first on carbon isotopes (panels D and E of Figs. 5.9 and 5.10), RPFR values converge to ±3‰ at low (25°C) temperatures and to less than 1‰ above 300°C once the MP2 level and triple-$\zeta$ basis set size are reached. These differences are similar in magnitude to the disagreement between the DBO-corrected PIMC calculations and experimental values (~0.56‰), suggesting that disagreements of this size could be due to inaccuracies in the reference potential energy surfaces.

Turning to hydrogen isotopes, RPFR calculations for $H_2$ converge to ±3‰ at triple-$\zeta$ (recall that CCSD treatment is exact for $H_2$). In contrast, for $H_2O$ and $CH_4$ we observe changes of up to 35‰ for $H_2O$ and 20‰ for $CH_4$ at 0°C in calculated RPFRs from CCSD to CSSD(T) (panels 5.9B, 5.9C). There is better (<4‰ difference) agreement between the CCSD(T) and reference calculations for $H_2O$ and $CH_4$. However, the reference calculations for both molecules are also done at the CCSD(T) level, though for the case of water, icMRCI and experimental data are used as well — regardless these similarities may result in some of the agreement seen between at the CCSD(T) level and reference results. Basis set size does not influence $CH_4$

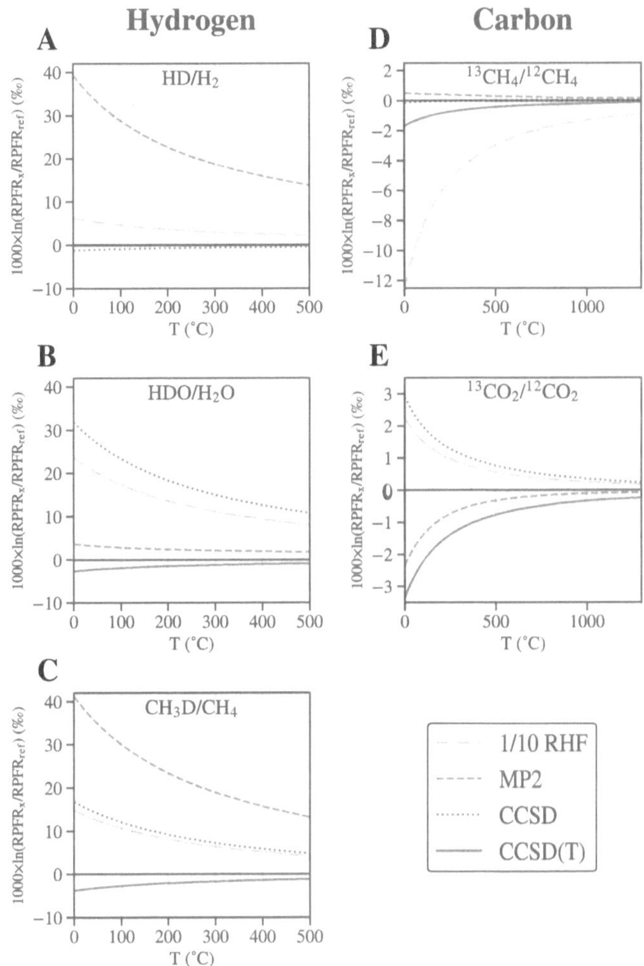



Figure 8: Changes in deuterium (A-C) and $^{13}$C (D-E) RPFRs with increasing level of correlation treatment at fixed basis set size (aug-cc-pVQZ) within the harmonic approximation, relative to harmonic calculations using the reference potentials, denoted as 'ref', as described in Section 2.5.1, and given as 0 on the y-axis. As noted in Section 4.5.1, y-axis is 1000×ln(RPFR$_x$/RPFR$_{ref}$) in each case, where X is the level of correlation treatment used. Note that the RHF treatment is scaled down by a factor of 1/10. We note that CCSD(T) calculations are not shown in (A) as they are identical to CCSD for the

Figure 5.9: **Changes in deuterium (A-C) and $^{13}$C (D-E) RPFRs with increasing level of correlation treatment** at fixed basis set size (aug-cc-pVQZ) within the harmonic approximation, relative to harmonic calculations using the reference potentials, denoted as 'ref' as described in Section 5.3 and given as 0 on the y-axis. The y-axis is 1000×ln(RPFR$_X$/RPFR$_{ref}$) in each case, where X is the level of correlation treatment used. Note that the RHF treatment is scaled down by a factor of 1/10.

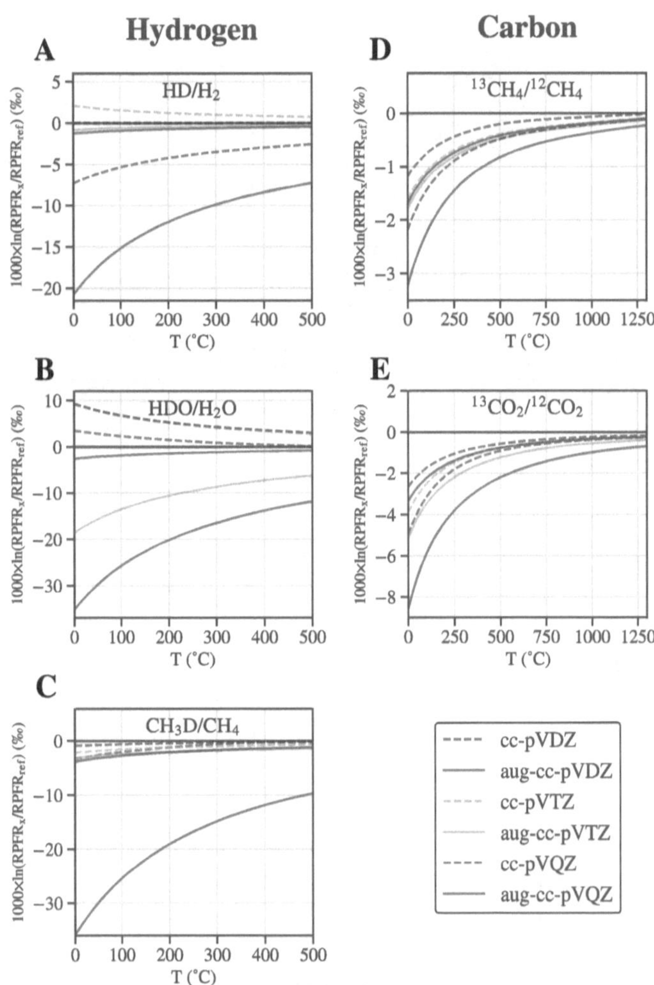


2380
2381 Figure 9: Changes in deuterium (A-C) and $^{13}$C (D-E) RPFRs with increasing basis set size at the same
2382 level of theory (CCSD(T)) within the harmonic approximation, relative to harmonic calculations using the
2383 reference potentials, denoted as 'ref' as described in Section 2.5.1 and given as 0 on the y-axis. As noted
2384 in Section 4.5.1, y-axis is 1000×ln(RPFR$_X$/RPFR$_{ref}$) in each case, where X is the basis set used.

Figure 5.10. **Changes in deuterium (A-C) and $^{13}$C (D-E) RPFRs with increasing basis set size** at the CCSD(T) level of theory within the harmonic approximation, relative to harmonic calculations using the reference potentials, denoted as "ref" as described in Section 5.3 and given as 0 on the y-axis. The y-axis is $1000 \times \ln(\text{RPFR}_X/\text{RPFR}_{ref}$ in each case, where X is the basis set used.

results (panel 5.10C) by more than 3‰ at triple-$\zeta$ or above. In contrast, for water (panel 5.10B), increasing from cc-pVQZ to aug-cc-pVQZ results in changes of up to 12‰ at 0°C. These results indicate that for $CH_4$ and $H_2O$, the calculations at the CCSD(T) level and cc-pVQZ have not converged to insignificant (sub per mil) levels. However, without still higher-level electronic structure calculations, precise estimates of errors are difficult to make for $CH_4$ and $H_2O$.

**Influence of anharmonicity and other quantum effects on RPFRs**

Here we discuss the importance of the inclusion of anharmonic and quantum effects included in RPFRs calculated with PIMC. Figure 5.11 compares calculations of $1000 \times \ln {}^D\alpha^{eq}_{CH_4(g)-H_2(g)}$, $1000 \times \ln {}^D\alpha^{eq}_{CH_4(g)-H_2O(g)}$, $1000 \times \ln {}^D\alpha^{eq}_{H_2O(g)-H_2(g)}$ (panel A), and $1000 \times \ln {}^{13}\alpha^{eq}_{CO_2(g)-CH_4(g)}$ (panel B) based on RPFRs calculated using the PIMC method to the harmonic calculations based on the same potential energy surfaces (reference harmonic line). Differences between $1000 \times \ln {}^D\alpha^{eq}_{CH_4(g)-H_2(g)}$, $1000 \times \ln {}^D\alpha^{eq}_{CH_4(g)-H_2O(g)}$, $1000 \times \ln {}^D\alpha^{eq}_{H_2O(g)-H_2(g)}$ calculated using PIMC vs. harmonic calculations are up to +10, -40, and -30‰ respectively, at 0°C (where the largest difference occurs). The same patterns (but smaller magnitude differences) are observed for carbon isotopic equilibrium, i.e., errors in RPFRs of different species do not cancel out (Fig. 5.11B). The difference in $1000 \times \ln {}^{13}\alpha^{eq}_{CO_2(g)-CH_4(g)}$ for PIMC vs. harmonic calculations is up to -2.5‰ at 0°C.

[86] and [98] demonstrated that the equilibrium methane clumped isotopic compositions of $^{13}CH_3D$ do not change by more than 0.06‰ from 27 to 327°C when PIMC or harmonic partition function ratios are used. This is due to almost perfect cancellation of nearly identical errors in harmonic partition function ratios of clumped methane isotopologues as we discussed in Chapter 4. However, when calculating fractionation factors between different species, the error of the harmonic partition function ratios only cancels partially (i.e., Fig. 5.11A).

This shows that unlike the case for clumped isotope studies, the cancellation of errors in harmonic partition function ratios or PIMC approaches cannot be assumed when dealing with equilibrium isotope-exchange reactions between different species. For example, neglecting anharmonic and quantum effects accounted for by PIMC leads to significant (up to ~40‰) errors in calculated $1000 \times \ln {}^D\alpha^{eq}$ values for hydrogen isotopes and ~2.5‰ errors in theoretical $1000 \times \ln {}^{13}\alpha^{eq}$ values for carbon isotopes for environmentally relevant conditions on Earth surface (i.e., at or above 0°C) for the species examined here.

**Diagonal Born-Oppenheimer correction**

The effect of the DBO correction on the calculated magnitudes of isotopic fractionations of D/H between small molecules has been explored in previous calculations,[105,174,179,190] and compared to experiments for $HD - H_2O$ [184] and $D_2 - HCl$ [191]. To assess whether or not the DBO correction is of importance to an ex-

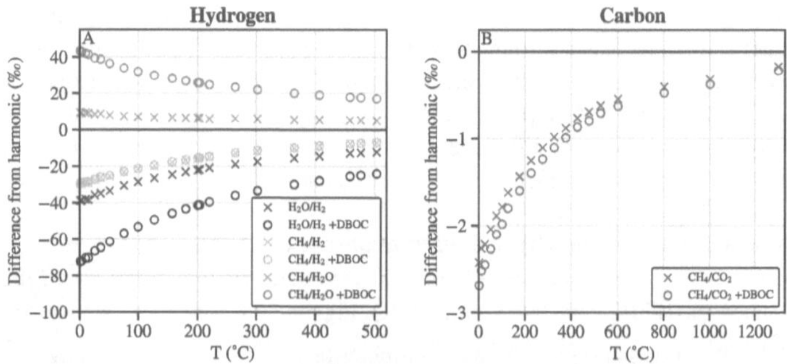



**Figure 5.11: Differences between PIMC calculations, for non-DBO-corrected and DBO-corrected cases (given as "+DBOC")** (A) for $1000 \times \ln{}^D\alpha_{CH_4(g)-H_2(g)}$, $1000 \times \ln{}^D\alpha_{CH_4(g)-H_2O(g)}$, and $1000 \times \ln{}^D\alpha_{H_2O(g)-H_2(g)}$ relative to corresponding harmonic ("reference harmonic") calculations without DBOC. Difference calculated as $1000 \times \ln{}^D\alpha_X$ minus $1000 \times \ln{}^D\alpha_{harmonic}$. (B) for $1000 \times \ln{}^{13}\alpha_{CH_4(g)-CO_2(g)}$ relative to corresponding $1000 \times \ln{}^{13}\alpha_{CH_4(g)-CO_2(g)}$ relative to harmonic calculations without DBOC. In all cases, the same potential energy surface is employed, as described Section 5.3

change reaction involving some atom A, the difference in electron density on A in two different molecules must be considered in addition to the mass of A. The DBO correction will increase in magnitude as (i) A decreases in mass (i.e., a bigger correction for hydrogen vs. carbon); (ii) A is substituted by a heavier isotope (i.e., a bigger correction for T/H vs. D/H, where $T \equiv {}^3H$ is tritium); and (iii) A is bonded to electron-withdrawing atoms or groups in one molecule and electron-donating in the other (e.g., the correction for $H_2O/H_2$ is greater than for $CH_4/H_2$). These empirical rules follow from the expression for DBO energy correction.[172]

Our results follow these guidelines. In particular, we find that DBO corrections have magnitudes of up to 34‰ on calculated hydrogen fractions for $H_2O - H_2$ and $CH_4 - H_2O$ but are negligible (<0.5‰) for calculations for hydrogen isotope fractionations for $CH_4 - H_2$ and carbon isotope equilibrium for $CH_4 - CO_2$ (see Fig. 5.11A and B). [90] did not employ DBO corrections as they considered them minor for systems exchanging atoms other than hydrogen; for exchange of hydrogen isotopes, they argue, the corrections computed at the time[190] were poorly constrained, such that the error on the correction approached the size of the correction. Using modern electronic structure theory methods, DBO corrections have been recomputed[174] with better accuracy. Importantly, the application of these more accurate DBO corrections reduces the difference of theoretically derived $1000 \times \ln{}^D\alpha_{H_2O(l)-H_2(g)}$ values from 24.5‰ to 0.49‰ relative to experimental determinations, and from 0.66‰ to 0.56‰ for $1000 \times \ln{}^{13}\alpha_{CH_4(g)-CO_2(g)}$ values. This agreement would indicate that their inclusion does yield more accurate calculation, especially for hydrogen isotopes.

We note that to our knowledge, this work presents the first case in which DBO corrections have been included in a path-integral statistical mechanical calculation. Non-Born-Oppenheimer effects are typically discussed in the context of non-adiabatic dynamics. The necessity to correct the path-integral result for such a small energy difference highlights two important aspects of the equilibrium isotope effect: (i) some gas mixtures have DBO corrections up to 6-8 cm$^{-1}$, large enough to confirm the role of these non-Born-Oppenheimer effects on the equilibrium distribution experimentally; (ii) path-integral calculations using the best available potential energy surfaces are accurate enough to reveal such small energy differences between the electronic ground states of small molecules.

**Differences in theoretical vs. experimental determinations of equilibrium fractionation factors**

The sensitivity tests and comparisons described in the previous sections indicate the following: (i) inclusion of anharmonic and quantum effects via use of the PIMC approach has a significant (up to 40‰) effect on equilibrium $1000 \times \ln{}^{D}\alpha^{eq}$ values for $1000 \times \ln{}^{D}\alpha_{CH_4(g)-H_2(g)}$, $1000 \times \ln{}^{D}\alpha_{CH_4(g)-H_2O(g)}$, and $1000 \times \ln{}^{D}\alpha_{H_2O(g)-H_2(g)}$ from 0 to 500°C. (ii) DBO corrections further change the computed RPFRs by up to 34‰ for $1000 \times \ln{}^{D}\alpha_{CH_4(g)-H_2O(g)}$ and up to -34‰ for $1000 \times \ln{}^{D}\alpha_{H_2O(g)-H_2(g)}$. (iii) RPFRs computed with PIMC for $^{13}$C are likely accurate to at least ±3‰; the inaccuracy appears to be predominantly due to the electronic structure calculations (i.e., the quality of the PES). (iv) The accuracy bounds for the RPFRs involving hydrogen isotopes are more difficult to estimate but are likely <10‰ based on calculations presented in Figures 5.9 and 5.10 and are likewise related to the approximations employed when calculating the potential energy surfaces. Higher-level calculation will be needed to resolve this.

We observe that when RPFRs for hydrogen isotopes are calculated using PIMC and the DBO corrections, we achieve close agreement with experimental observations of ${}^{D}\alpha^{eq}_{H_2O(l)-H_2(g)}$ (Fig 5.3) and ${}^{D}\alpha^{eq}_{CH_4(g)-H_2(g)}$ (Fig. 5.7). In contrast, not including these terms in the computation of ${}^{D}\alpha^{eq}_{H_2O(l)-H_2(g)}$ results in offsets in calculation relative to experimental observations of ~25‰. This indicates that these corrections, which are not always included in calculation of RPFRs, can be important for hydrogen isotope equilibrium calculations. We note that in making this comparison between theory and experiment we assume that the experiments are of sufficient quality both in precision and accuracy to reflect the true equilibrium isotopic compositions. We believe that our equilibrations of CH$_4$ and H$_2$ given the bracketing approach are sufficiently accurate and precise for this purpose. This is because the residual of the data to our best-fit line is ±5.9‰ ($1\sigma$) vs. whereas differences between theoretical treatments can be greater (up to 30‰ for using harmonic calculations with our reference potential). In contrast, for water-H$_2$ hydrogen-isotope equilibrations, experimental results show scatter of >25‰ at given temperatures. This scatter makes exact comparisons between experiment and theory more challenging and indicates that future experimental equilibrations of $H_2(g)$ and $H_2O(l)$ using modern techniques will be useful in comparison of experiments and theory.

## 5.5 Summary

Experimentally interpolatable calibrations of hydrogen isotopic equilibrium between methane and liquid water as a function of temperature prior to this work were only available for temperatures above 200°C.[140] Additionally, calculation of $CH_4 - H_2O$ hydrogen isotopic equilibrium from 0 to 200°C differ in value by $\sim 160‰$ between each other and the extrapolation of the experimental calibration of [140] to low temperatures. Here we presented an experimental calibration of hydrogen isotopic equilibrium $CH_4 - H_2O(l)$ that is interpolatable from 3 to 200°C. This was done by equilibrating the hydrogen isotopes of $CH_4$ and $H_2$ using $\gamma$-$Al_2O_3$ as a catalyst based on a bracketing approach and combining this calibration with previous experimental determinations of hydrogen isotopic equilibrium between molecular hydrogen and water. We then compared this work both to new calculation of equilibrium hydrogen isotopic fractionation factors in the system $CH_4-H_2-H_2O$ using Path Integral Monte Carlo (PIMC) calculations.

We found that our experimental calibration of $1000 \times \ln{}^D\alpha_{CH_4(g)-H_2(g)}$ agrees 1:1 with calculation performed using the PIMC approach (with or without the DBO correction). Additionally, comparison of previous experimental determinations of $1000 \times \ln{}^D\alpha_{H_2O(l)-H_2(g)}$ agree 1:1 within 1 s.e. with our theoretical DBO-corrected PIMC calculations. We investigated potential sources of error for the calculation. It appears that deviations of at least $\sim 10‰$ for calculations of hydrogen isotope equilibrium between species are plausible given changes in both the theoretical level and basis set sizes used to calculate the potential energy surface for $H_2O$ and $CH_4$, which are then used for the PIMC calculations; these errors are challenging to exactly quantify due to a lack of convergence with increasing level of theoretical treatment. We note that anharmonicity, quantum effects, and DBO corrections all can have large effects (up to 34‰ for hydrogen isotopes) on final calculated RPFRs and $\alpha^{eq}$ values and their inclusion was needed here to yield agreement between theory and experiment. Finally, we additionally provided a theoretical calibration of $1000 \times \ln{}^{13}\alpha^{eq}_{CH_4(g)-CO_2(g)}$ based on the PIMC method. It agrees with experimental data from 200 to 1300°C with an average offset of 0.56‰.

# Chapter 6

# $D$ AND $^{13}C$ ISOTOPIC EQUILIBRIA IN ALKANES

## 6.1  Introduction

We explored the clumped isotope effects in methane in Chapter 4 and heavy isotope effects in methane and other small molecules in Chapter 5. Here we attempt to extend the same rigorous treatment to mid-sized molecules: ethane and propane. There are two main challenges: (i) obtaining an accurate potential energy surface and (ii) addressing a large number of isotopologues.

There is no high-quality potential energy surface available for ethane or propane in the literature, as constructing these surfaces based on hundreds of thousands of single point icMRCI or even CCSD(T) calculations extrapolated to CBS limit is currently beyond reach. Thus, we first must first validate the accuracy of the potential energy surfaces within reach and quantify (at least approximately) potential inaccuracies that result from their use. These are then compared to the anharmonic effects to ultimately conclude whether PIMC calculations are necessary and feasible to achieve the best accuracy.

The number of unique isotopologues grows rapidly with increasing the size of the molecule from 3 for molecular hydrogen to 36 for ethane and 216 for propane. Fortunately, at natural abundances only a few of those have measurable abundances (mainly singly and doubly substituted isotopologues). We do not attempt to calculate all isotopic equilibria with PIMC, since most of them have a negligible effect on the experimentally accessible abundances of isotopologues. The equilibrium constants that involve triple and more substitutions are therefore computed based on harmonic RPFR's and used in conjunction with the PIMC-based equilibrium constants.

## 6.2   Methods

We studied methane, ethane, propane, molecular hydrogen and water. The RPFRs of singly and doubly substituted isotopologues were calculated as described in section 3.3 using two distinct approaches: (i) harmonically using the approach outlined in [92, 96] and (ii) using the path integral Monte Carlo (PIMC) method following the approach developed by [102]. We also evaluate the RPFRs for all other isotopologues using only the harmonic approach.

Although RPFR calculations with either approach require knowledge of the potential energy of the molecule as a function of geometry, these must be done differently for the two approaches as discussed below. We compare the two approaches in detail throughout this work. Briefly, h armonic c alculations r equire l ess computational resources compared to the PIMC method. As a result, for a given amount of computational resources available, more accurate potentials can be used, or alternatively larger molecules can be calculated with the harmonic method relative to the PIMC method. However, the harmonic calculations make a number of approximations, that can lead to systematic errors in calculated RPFRs. In some cases, these errors mostly cancel out when harmonic RPFRs are used to calculate observables like the equilibrium constants,[86,98] but this is not always case.[86,99]

### PIMC calculations

The path integral Monte Carlo (PIMC) method allows us to calculate the RPFRs without relying on the validity of the harmonic approximation. In the limit of infinite n umber o f b eads $n$, t he p artition f unction o f t he r eal q uantum s ystem is equal to the partition function of the fictitious n-times larger c lassical s ystem (see Eq. 1.5). In practice we always use approximations to evaluate the potential energy and sufficiently large but finite $n$ (Table 6.1 is cho sen. The direct scaled-coordinate estimator[102] was used to calculate the RPFR for every single heavy substitution relative to the lighter isotopologue and for every double substitution relative to the corresponding singly-substituted isotopologue. For each pair, the heavier of the

| Molecule | Number of beads at this temperature | | | | | | | | | |
|---|---|---|---|---|---|---|---|---|---|---|
| | 0°C | 25°C | 50°C | 75°C | 100°C | 150°C | 200°C | 300°C | 400°C | 500°C |
| $H_2$, $H_2O$ | 950 | 865 | 781 | 697 | 613 | 529 | 445 | 361 | 277 | 193 |
| alkanes | 420 | 383 | 355 | 332 | 313 | 275 | 253 | 214 | 196 | 166 |

Table 6.1: Number of beads employed in the Path-Integral calculations. Water and dihydrogen have higher frequency vibrations compared to alkanes. Therefore, alkanes require fewer beads to achieve the target accuracy.

| Molecule | $H_2$ | $H_2O$ | Methane | Ethane | Propane |
|---|---|---|---|---|---|
| D | 430-500 | 80-110 | 100-110 | 69-70 | 18-60 |
| D + D | 430-500 | 100-110 | 60-70 | 23-34 | 9-30 |
| $^{13}C$ | | | 20-25 | 20-21 | 10-20 |
| $^{13}C$ + D | | | 78-80 | 24-30 | 19-30 |
| $^{13}C$ + $^{13}C$ | | | | 10-11 | 10 |

Table 6.2: Number of Path-integral samples (in millions) that the results were averaged over. Since the samples were drawn every 10 steps, the number of Monte-Carlo steps is 10 times larger. For propane, the larger ranges are because there is only one secondary (center) carbon and two primary (terminal) carbons and correspondingly 2 vs 6 deuterium atoms. Thus, the terminal carbons are sampled twice as often and the center carbon and the terminal hydrogens are sampled three times as often as the center hydrogens.

two isotopologues was sampled with PIMC in Cartesian coordinates with hundreds of millions of Monte Carlo steps (see Table 6.2). Each Monte Carlo step consists of (1) moving the entire ring-polymers by a small random displacement in each coordinate and (2) $P/j$ staging moves (rounded up to the nearest integer).[31] The average displacement and staging length $j$ were set such that $40\pm2\%$ of all proposed staging moves are accepted to optimize sampling efficiency. Prior to any data collection, each sampling trajectory was equilibrated for $10^5$ Monte Carlo steps. Thereafter, ring-polymer configurations were sampled every 10 Monte Carlo steps. The number of samples collected for each isotopologue is given in Table 6.2.

**Molecular potentials for the harmonic calculations**

We calculate harmonic RPFR's using Eq. (3.25). The harmonic frequencies are calculated using the ORCA 5.2 program[192–194] and the MOLPRO program.[195] All ORCA calculations are performed with the most stringent default option "very tight" for the convergence criteria on the finest default grid (DEFGRID3). An example input file and the brief description of calculation algorithm are provided in the Appendix A.1.

The potential energy of molecules as a function of their geometry at present cannot be obtained exactly for any but the smallest of molecules (e.g., molecular hydrogen). Therefore, we use various approximations to calculate the potential energy that are available within ORCA. We have calculated harmonic frequencies using a variety of common methods employed in prior isotope geochemistry work to assess the relative importance of the accurate description of the potential energy of the molecule.These include density functional theory (DFT) with the B3LYP functional[196] and four increasingly large Pople's[197] basis sets (6-31G, 6-311G, 6-311G(d,p), 6-311++G(d,p)) as well as second order Møller-Plesset perturbation theory (MP2), coupled-cluster with single and double excitations (CCSD), and the traditional "gold standard" CCSD(T) where triple excitations are included perturbatively. We use the correlation-consistent basis sets[171] that are designed for post-Hartree-Fock calculations. We have used four basis sets of increasing size: cc-pVDZ, aug-cc-pVDZ, cc-pVTZ, and aug-cc-pVTZ.

For completeness, we also include calculations using the restricted Hartree-Fock (RHF) method. This mean-field theory does not take into account the electron-electron correlations and, as such, is generally not used in modern calculations. Finally, we also calculate the harmonic frequencies using two empirical force fields (CHARMM and AIREBO) since there are calculations for isotope effects considered in this book based on both of these force fields.[86,102]

**Molecular potentials for the PIMC calculations**

PIMC calculations of the RPFR's presented require the potential energy to be calculated for $> 10^{10}$ different molecular geometries for each isotopologue. In contrast, harmonic calculations described above only require about $10^3$ potential energy evaluations to compute the harmonic frequencies of propane and even fewer for the smaller molecules. The most cost-effective way to perform orders of magnitude more potential energy evaluations for PIMC is therefore different from the direct (i.e., geometry in, energy out) calculations that were done for the harmonic calculations.

Instead, we fit the potential energy surface (PES) of the molecule and use this fit in our PIMC calculations. This strategy enables us to use PI-based approaches for the molecules considered here with highly accurate PESs. That said, fitting the PES in sufficient detail still requires many more potential energy evaluations than the harmonic frequency calculations, limiting the use of PIMC with highly accurate PESs to relatively small and highly symmetric molecules. Less accurate (and therefore less computationally demanding) approximations of molecular potential can be used in conjunction with PIMC (or PIMD), but there is no *a priori* reason to expect these RPFRs to be more accurate than the (computationally less expensive) harmonic results with the same potential.

In our prior studies[98,99] we used what we considered the best potential energy surfaces for each molecule at the time ($H_2(g)$, $H_2O(g)$, $CO_2$ $(g)$, and $CH_4(g)$). Other than for $H_2$ these PESs used were calculated and published by others. We assume that the increasing the accuracy of the potential will in turn increase the accuracy of the calculated RPFR's of individual molecules. A possible issue with this approach taken in our prior work is that the methods used to calculate the PES of different molecules relied on different approximations to obtain the molecular potential. As such, when RPFRs are used to find the fractionation factor between different molecules (Eq. 3.21), errors associated with different methodologies may not cancel resulting in inaccurate fractionation factors. We have observed this for the harmonic RPFRs both in this study and in [99] — the RPFRs computed with different methods and basis set sizes differ by 10s to 100s per mil. However, when the relevant ratios of RPFRs (Eq. 3.21) are calculated to obtain a fractionation factor, most of these differences cancel out. As such, the resultant fractionation factors then show much smaller errors when all molecules are treated within a single method.

Therefore, we take a different approach here and fit the PES for each molecule ourselves in a consistent manner. The potential energy is obtained using CCSD(T)-F12A (explicitly correlated coupled-cluster with single, double, and perturbative triple excitations) method[198] with aug-cc-pVTZ basis set.[171] All potential energy calculations for the PES fitting were performed within MOLPRO 2021.3 package.[195] We note that as this method is implemented within MOLPRO, it can also be readily used by others. We will refer to this method using a shorthand "F12/ATZ" throughout this Chapter.

The "explicitly correlated" descriptor in the name of the method highlights that in contrast to the standard CCSD(T) algorithm, the 2-electron wavefunctions are added to describe the electron correlation at short distances. This is commonly termed the "cusp condition."[199] Without such functions, this portion of electron correlation converges slowly as the basis set size increases. Direct comparisons have shown that the explicitly correlated CCSD(T)-F12A with aug-cc-pVTZ basis set achieves comparable accuracy to CCSD(T) with the larger aug-cc-pV5Z.

Since molecular hydrogen is a diatomic molecule, its potential energy depends only on one coordinate – the distance between two hydrogen atoms; thus, we fit the PES for dihydrogen using a 1-dimensional spline interpolation on 76 F12/ATZ calculations for interatomic distances between 0.5 and 6 Bohr (equilibrium geometry is at 1.402 Bohr). Such direct fitting on a dense grid becomes prohibitively expensive for larger molecules as the number of coordinates grows rapidly (3 for water, 10 for methane, 16 for ethane and 22 for propane). A similar interpolation for propane would require $76^{22} \approx 10^{41}$ potential energy evaluations which is not possible with current computing constraints.

Instead, we fit delta-learned permutationally invariant potential energy surfaces to water, methane, ethane, and propane.[200] The permutational invariance with respect to identical atoms is used to reduce the number of distinct arrangements of atoms by not double counting symmetric orientations. This reduces the number of points needed for the fit to be accurate.[201] The delta learning approach is used to construct a fit to F12/ATZ-quality surface for each molecule at reduced computational cost. Delta learning works by separating the problem into two separate fits. The first fit contains most of the complexity of the PES, and thus requires a large number

| Molecule | # of geometries | | RMSE (cm$^{-1}$) | | |
|---|---|---|---|---|---|
| | low level | high level | low level | high level | total |
| Water | 9902 | 2401 | 1.35 | 1.58 | 2.08 |
| Methane | 28948 | 3000 | 1.71 | 1.18 | 2.08 |
| Ethane | 33522 | 2655 | 4.52 | 2.32 | 5.08 |
| Propane | 19921 | 4990 | 4.82 | 1.73 | 5.12 |

Table 6.3: Number of single-point energy evaluations for fitting the delta-learned PES and the corresponding root-mean-square error of the fit.

of geometries (order 10,000 for the molecules and target accuracy in this study). The "low level" theory is used to sample the geometries in this step. Although the resulting PES does not have the desired energy accuracy, this is addressed by the second fit. Energy is calculated with the "high level" theory for a subset ($\sim$10%) of the molecular geometries from the first fit and the difference in energy between the "low level" and the "high level" theory is fitted. This delta learning approach cuts the computational cost by about an order of magnitude. We use B3LYP/6-311+G(d,p) DFT and CCSD(T)-F12A/aug-cc-pVTZ as the "low level" and "high level" methods, respectively. The RMSE, describing the quality of each fit is 2 cm$^{-1}$ for water and methane and 5 cm$^{-1}$ for ethane and propane (see Table 6.3).

These PESs were fitted for use in the PIMC calculations. However, we also want to be able to compare directly PIMC and harmonic calculations in order to be able to evaluate how inclusion of anharmonic terms changes final calculated fractionation factors vs. using the harmonic approximation. To that end, we calculate the harmonic frequencies based on the fitted PESs using FORTRAN subroutines from the Numerical recipes text.[54]

**The diagonal Born-Oppenheimer correction**

We explore the importance of the DBO corrections on the molecules studied here using the CFOUR package[202] to calculate the DBO correction at the CCSD level of theory and aug-cc-pVTZ basis set. The exact DBO correction depends on both the molecular geometry and the level of theory. However, we find that the dependence is weak and therefore we compute these corrections at CCSD/ATZ level of theory at the equilibrium geometry of interest and approximate it is constant (independent of molecular geometry). With this assumption, the calculated DBO correction (see Table 6.4) affects the RPFRs via a free energy shift:[105]

$$\text{RPFR}_i^{\text{DBOC}} = \text{RPFR}_i \times \exp\left(-\frac{\Delta C}{k_B T}\right) \tag{6.1}$$

We compared the RPFRs calculated (i) with no DBO correction; (ii) with the approximate (given in Eq. 6.1) and (iii) with the exact DBO correction for dihydrogen. At 25°C, the approximate DBO correction recovers a 6‰ decrease in RPFR, while

| Isotope | Hydrogen | Water | Methane | Ethane | Prop. (term.) | Prop. (center) |
|---|---|---|---|---|---|---|
| D | 1.235 | 7.511 | 1.109 | 0.364 | 0.444 | -0.210 |
| $^{13}$C | - | - | 0.703 | 0.708 | 0.694 | 0.733 |

Table 6.4: DBO corrections $\Delta C$ (cm$^{-1}$) for single heavy atom substitution of a given molecule relative to free atom. All corrections are calculated with the CFOUR package at CCSD level of theory.

the exact DBO correction (including its dependence on the H-H distance) yields a 7.2‰ decrease. Due to limited computational resources, we do not test the validity of this approximation in larger molecules and assume it brings the RPFR substantially closer to the true value just like for dihydrogen. Finally, we note that the DBO corrections do not affect the clumped isotope equilibria.[179]

**Validation of the reference molecular potential**

To estimate the errors from approximations to potential energy, we compare the RPFRs computed in this study and in our previous study[99] to the best *ab initio* approximations to Born-Oppenheimer potentials to date for dihydrogen,[203,204] water[205] and methane.[206] For dihydrogen we are also able to estimate the effect of approximating the DBO correction as being independent of the molecular geometry using the DBO correction surface.[207]

The errors in RPFRs relative to the best potentials are shown in Table 6.5. We show the difference in the PES for dihydrogen relative to the best PES available to date[203,204] on Fig. 6.1. Because only the relative energies are significant, all three surfaces are shifted so that the origin (0,0) corresponds to the minimum energy and the corresponding equilibrium H-H distance.

The CCSD/cc-pVQZ PES we used in our previous work[99] is shown on Figure 6.1 in orange. The difference in harmonic frequency between this PES and our reference (current best) is very small, about 1 cm$^{-1}$. Recall that the harmonic frequency is a measure of the second (i.e., lowest, most important) order effect of the energy on RPFR. Visually, this manifests in a cubic (third order) shape of the orange deviation curve around 0 on Figure 6.1. Thus, the harmonic RPFR computed on this PES is only ~0.5‰ different from the best result (see Table 6.5).

The PI calculation picks up on the differences past second order, which in this case partially cancel the harmonic effect resulting in a smaller total difference. The F12/ATZ surface has about 3cm$^{-1}$ difference in harmonic frequency, resulting in parabolic shape of the green curve on Figure 6.1 and a larger (~1‰) error in harmonic RPFR at 25°C. On the other hand, the anharmonic effects in F12/ATZ potential are pretty much identical to the best PES. The F12/ATZ surfaces describe the anharmonic effects of the best *ab initio* PES well and yield harmonic RPFR's that are consistently underestimated computed with the best ab initio PES. In some cases, the harmonic and anharmonic discrepancy partially cancel each other, result-

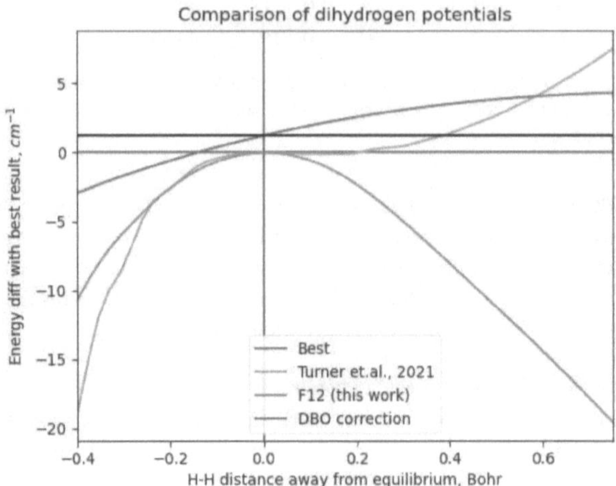


Figure 6.1: **Difference in the potential energy for molecular hydrogen relative to the best PES** (Pachucki, 2010; Pachucki and Komasa, 2014). The range of H-H distance (-0.4 to 0.75 Bohr away from equilibrium distance) covers potential energy of up to 10,000 $cm^{-1}$.

ing in a superficially good agreement with the best *ab initio* result. Overall, the F12/ATZ PES's used in this study are comparable to the surfaces we used in our prior work. Crucially, we can see that in all cases the RPFR's are underestimated using the F12/ATZ PES, resulting in partial error cancellation when experimentally observable quantities are computed. We expect this to be the case for ethane and propane as well, although at present we cannot test that, since the reference-quality PESs have not yet been developed for the molecules of that size. To maintain internal consistency, we will use the F12 surfaces as reference PES in each case even though we have access to the (essentially exact) published PES's for dihydrogen and water as well as a highly accurate methane PES.

The dependence of DBO correction on molecular geometry for dihydrogen is shown in Figure 6.1 in red and causes a 7.2‰ decrease in both RPFR(HD) and RPFR(D$_2$/HD) at 25°C. To make the calculations feasible, we must approximate the DBOC and the simplest approximation (assuming the DBOC is constant with respect to molecular geometry and given by its value at equilibrium geometry) recovers 6.0‰ decrease in RPFR for dihydrogen. We cannot rigorously test this for other molecules, but we expect it to be a similarly good approximation for all molecules considered here.

Table 6.5 highlights the difficulties in computing RPFR's to high precision for any but the smallest molecules. The RPFR's are sensitive to the potential energy of the

| $\delta$ RPFR, ‰ | Harmonic | | Anharmonicity | | Total | |
|---|---|---|---|---|---|---|
| | [99] | F12/ATZ | [99] | F12/ATZ | [99] | F12/ATZ |
| HD | -0.41 | -1.05 | 0.22 | -0.35 | -0.19 | -1.15 |
| $D_2$ / HD | -0.49 | -1.24 | 0.25 | -0.34 | -0.24 | -1.39 |
| HDO | 0.41 | -2.74 | 0.07 | 2.57 | 0.49 | -0.17 |
| $D_2O$ / HDO | 0.48 | -3 | 0.12 | 2.69 | 0.6 | -0.31 |
| $^{13}CH_4$ | -0.23 | -0.21 | 0.27 | 0.02 | 0.05 | -0.19 |
| $CH_3D$ | -4.59 | -3.39 | 3.23 | -0.98 | -1.36 | -4.37 |
| $CH_2D_2$ / $CH_3D$ | -4.62 | -3.37 | 2.63 | -1.65 | -1.99 | -5.02 |
| $^{13}CH_3D$ / $^{13}CH_4$ | -4.61 | -3.4 | 3.24 | -1.16 | -1.36 | -4.57 |

Table 6.5: Difference in RPFR (‰) at 25°C relative to the best published potential energy surfaces: [203, 204] for dihydrogen, [205] for water and [206] for methane.

molecule, which at present cannot be obtained with sufficient accuracy to achieve sub-per-mil precision in RPFR's involving D exchange. This is a general issue for any method, that aims at highly accurate RPFR's: PIMC,[86,93,98,99] PIMD,[180,208,209] and methods to include anharmonic effects based on perturbation theory.[100,130,174]

Table 6.5 also presents a cautionary tale that our error estimates due to use of approximate potentials are only taking into account the second order (harmonic) effects. At least for the small molecules considered here, these errors are comparable with those arising from the higher order (anharmonic) effects in the PES, meaning that the true error can be larger or smaller than the harmonic error contribution depending on whether the two parts partially cancel or reinforce each other. This analysis lets us estimate the errors for the quantities we present in this study. We expect the fractionation factors involving D and carbon-13 to be accurate to 5‰ and ~0.5‰ respectively, at 25°C. We expect similar errors in the position-specific isotope effects. Finally, the D + D and $^{13}$C + D clumped isotope effects should be accurate to ~0.5‰ and ~0.1‰ respectively. These error estimates are given for 25°C and should be scaled up (down) for higher (lower) temperatures within the range covered in this study.

## 6.3 Results

Here we present the results of the RPFR, fractionation factor, and clumped isotope calculations for the for the five molecules studied based on the different electronic structure methods for harmonic calculations as well a comparison of PIMC vs. harmonic calculations for the reference PES (F12/ATZ). We organize the results around the key factors that we wish to understand and discuss: (i) the range of potential differences associated with choice of PES for harmonic calculations; (ii) the size of PIMC corrections on harmonic calculations; And (iii) the importance of DBO corrections. For all three cases, we examine whether differences in RPFRs calculated using different methodologies do or do not cancel when RPFRs are combined to calculate fractionation factors.

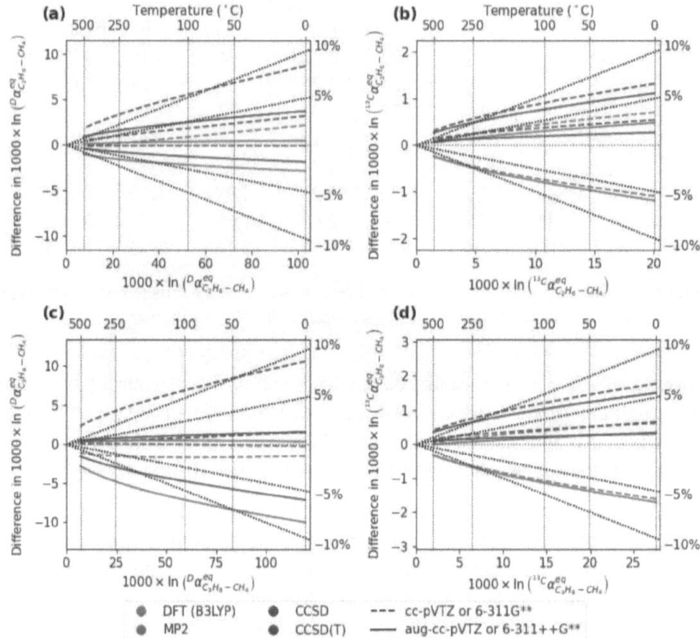

Figure 6.2: **Difference in harmonic fractionation of carbon-13 (b,d) and deuterium (a,c) with methane for ethane (a-b) and propane (c-d)** computed with 4 commonly used electronic structure methods and triple-$\zeta$ basis sets relative to the F12/ATZ method. Dotted vertical lines label temperatures (from 0°C to 500°C). The slanted black lines denote relative difference in fractionation (5 and 10%) and the horizontal line is placed at y=0 where the reference (F12/ATZ) result would be.

### Differences in harmonic calculations using different PESs

We first present results for bulk fractionation factors between the studied alkanes for methane vs. ethane in Fig. 6.2(a-b) and methane vs. propane in Fig. 6.2(c-d). This allows us to observe how well errors cancel when RPFRs based on the same method are combined to calculate a fractionation factor. We present differences on the y axis both in absolute magnitude and relative percent vs. the magnitude of the fractionation factor ($1000 \times \ln \alpha^{eq}$) on the x axis. We only present results for the larger basis set sizes (triple zeta and augmented triple zeta) and the more accurate electronic structure method (omitting Hartree-Fock and force field-based results here, since they are off the scale of these plots). All differences are given relative to the F12/ATZ treatment. This will be done through this section, and we wish to emphasize that this is not done to indicate this is best technique to date, but rather because we use this for the PIMC calculations. For both carbon and hydrogen we see up to about 10% relative difference in equilibrium fractionation $1000 \times \ln \alpha^{eq}$ between the studied methods. In general, absolute differences increase, while relative differences decrease at lower temperatures. We do not observe a convergence to similar values as the quality of the method improves from lowest quality of MP2 through B3LYP to CCSD, to CCSD(T) nor use of triple zeta vs augmented triple zeta. Rather, we see that different methods largely scatter in the $\pm 10\%$ range. This is also the case for the smaller basis sets (DZ and ADZ; Fig. B.2) indicating that for alkanes, a $\pm 5$ to 10% relative error for harmonic calculations is difficult to improve upon up to and including CCSD(T)/ATZ level of theory. We note, thought that use of computational less expensive descriptions of the potential energy of the molecules based on restricted Hatree-Fock and empirical force fields that we tested (AIREBO and CHARMM) yields much larger relative difference of about 20% for RHF and as much as 80% for the force fields (see Fig. B.1).

We next turn to fractionation factors involving molecules beyond alkanes including $H_2$ and $H_2O$. We provide a similar plot as Fig. 6.2 for all species relative to $H_2$ in Fig. 6.3(a-b) and of water relative to other molecules in Fig. 6.3(c-d) — for the latter we do not include the $H_2$ as the size of alpha for this is $\sim 10$ times larger than the others. All calculations are presented for the temperature range from $0°C$ to $500°C$, but temperatures are not given as they vary vs. $1000 \times \ln \alpha^{eq}$ depending on the molecule pair. We observe that fractionation factors relative to dihydrogen show absolute differences of up to $\pm 50\%_0$ relative to the F12/ATZ result. However, because the absolute values of $1000 \times \ln \alpha^{eq}$ are $\sim 10x$ larger than for the alkanes, relative differences are $\pm \sim 3\%$ vs. the 5%-10% for the alkanes. Fractionations relative to water appear the most sensitive (in the relative sense) to differences in how the potential energy was calculated with relative differences larger than 25% observed and absolute differences up to $\pm 40\%_0$. Fractionations involving water do not show the same monotonic increase in the absolute difference of fractionation factors with decreasing temperature. As was the case with alkanes, double zeta and augmented double zeta basis sets yield similar errors except for fractionations involving water, which differ significantly with 6-31G and 6-311G basis sets at the DFT (B3LYP) level of theory (see Fig. B.3). Less accurate RHF method shown in Fig. B.4 differs significantly from the results presented in Fig. 6.3.

Position-specific isotope effect in propane (Fig. 6.4) shows similar trends to the fractionation presented on Fig. 6.2. The relative differences in $1000 \times \ln \alpha^{eq}$ up to $\pm 10\%$ are observed for both deuterium and carbon-13. Analogously, DFT underestimates the isotopic preference for carbon-13, while CCSD overestimates it relative to the reference method. Smaller basis set size does not change these general observations (see Fig. B.5), while cheaper methods (RHF, AIREBO, CHARMM) yield results that differ significantly from the reference for deuterium, but not for carbon-13 (Fig. B.6). CHARMM force field gives position-specific deuterium isotope effect in propane in better agreement with the reference method as compared to the AIREBO

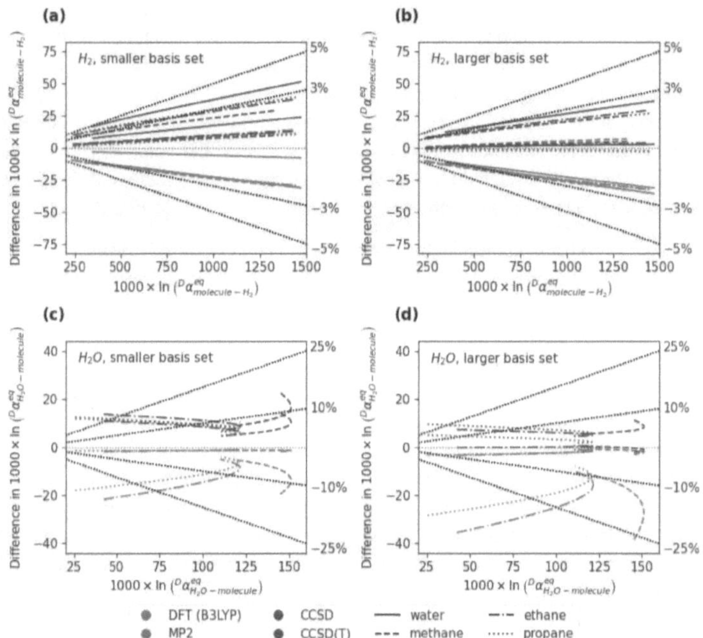


Figure 6.3: **Difference in harmonic fractionation of deuterium with dihydrogen (a-b) and water (c-d)** computed with 4 commonly used methods relative to the F12/ATZ method. 6-31G$^{**}$ was used for the DFT calculations and cc-pVTZ basis set for all other methods in panels (a,c); 6-311++G$^{**}$ was used for the DFT calculations and aug-cc-pVTZ basis sets for all other methods in panels (b,d). Dotted vertical lines label temperatures (from 0°C to 500°C). The slanted black lines denote relative difference in fractionation (3%, 5%, 10% and 25%) and the horizontal line is placed at y=0 where the reference (F12/ATZ) result would be.

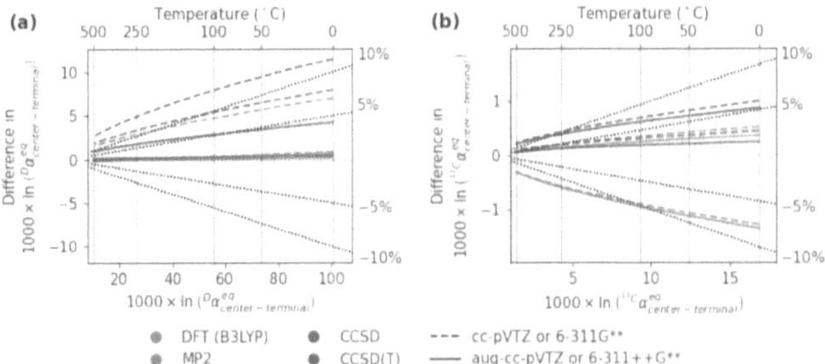

Figure 6.4: **Differences in position-specific isotope effect for (a) deuterium and (b) carbon-13** computed with four methods at the triple-$\zeta$ basis set size. All computations are harmonic relative to the Harmonic F12/ATZ. Dotted vertical lines label temperatures (from 0°C to 500°C). The slanted black lines denote relative difference in fractionation (3%, 5%, 10% and 25%) and the horizontal line is placed at y=0 where the reference (F12/ATZ) result would be.

force field. This is in contrast to the methane-propane fractionation, presented on Fig. B.1(c), where AIREBO agreed with the reference method better. We next turn to the clumped heavy isotope effect for the closest placement of two heavy atoms in each molecule, which produces the strongest effect (i.e., largest value of $\Delta^{eq}$). D+D clumped isotope effect shown in Fig. 6.5 deviates by less than 5% relative error as compared to the reference method (F12/ATZ). The range for $^{17}$O + D and $^{18}$O + D effects is even tighter (i.e., ±2%) and similarly for $^{13}$C + D effects, which are within 3% of the reference value (Fig. 6.6). $^{13}$C + $^{13}$C effects in ethane and propane deviate by as much as 8% relative to the reference (Fig. 6.7).

Just as with fractionation and position-specific effects, CCSD tends to overestimate clumping and DFT to underestimate it relative to the reference method. Again, there is not a definite convergence to the reference value with increasing the quality of the method past MP2. Empirical force fields (CHARMM, AIREBO) are inconsistent, giving excellent agreement (<10% relative error) with the reference value in some cases (see Fig. B.10(c-f) and Fig. B.11), but not in others (Fig. B.8(c-f)). In particular, AIREBO overestimates clumping in ethane and terminal carbon propane by more than 50%. RHF is more consistent relative to the reference result, typically overestimating it by about 10%. Just as for fractionation, using a smaller basis set does not dramatically shift the results further away from the reference value (Figs. B.7 and B.9). Interestingly, the basis set size is most important for the DFT calculations, where results obtained with the 6-31G and 6-311G basis set typically differ substantially more from the reference than those where additional polarization functions (i.e., 6-311G$^{**}$) and diffuse functions (i.e., 6-311++G$^{**}$) are included in the basis set (see Figs. B.7 and B.9).

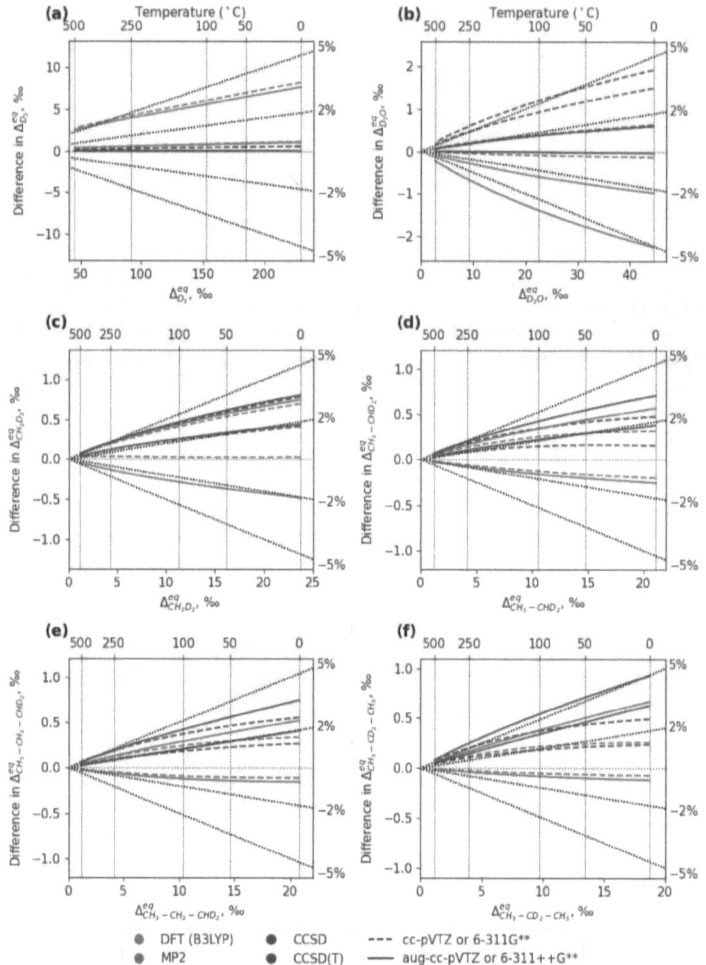



Figure 6.5: **Differences in** D + D **clumped heavy isotope effect** in (a) dihydrogen, (b) water, (c) methane, (d) ethane and (e-f) propane computed with four methods at four basis at the triple-$\zeta$ basis set size. All computations are harmonic relative to the Harmonic F12. Dotted vertical lines label temperatures (from 0°C to 500°C). The slanted black lines denote relative differences (2% and 5%) and the horizontal line is placed at y=0 where the reference (F12/ATZ) result would be.

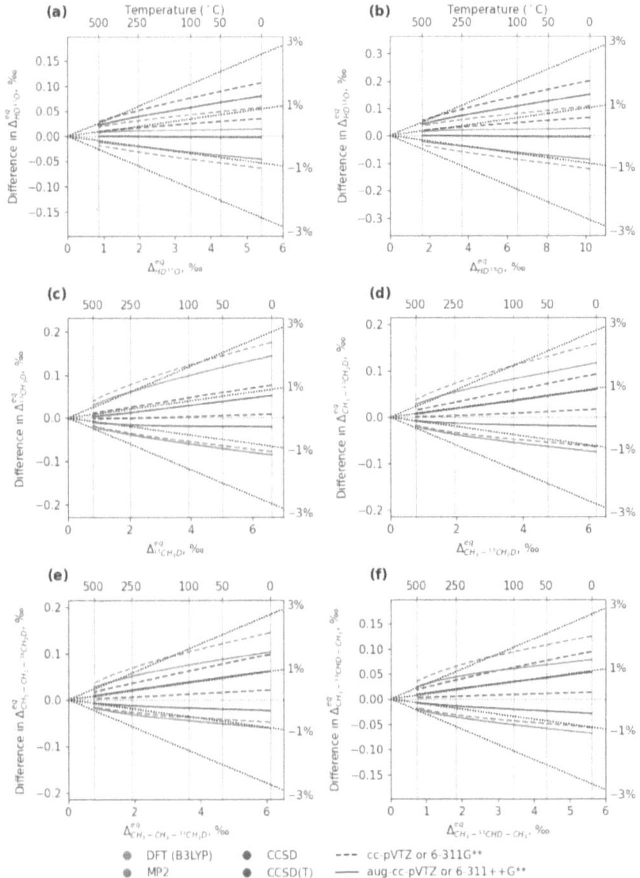

Figure 6.6: **Differences in clumped isotope effect involving one deuterium and one heavy atom (oxygen or carbon)** computed with four methods at four basis at the triple-$\zeta$ basis set size. The panels are as follows: $^{(17)}O + D$ (a) and $^{18}O + D$ (b) in water, $^{13}C + D$ in methane (c), ethane (d) and propane (e-f). All computations are harmonic relative to the Harmonic F12/ATZ. Dotted vertical lines label temperatures (from 0°C to 500°C). The slanted black lines denote relative difference in fractionation (1‰ and 3‰) and the horizontal line is placed at y=0 where the reference (F12/ATZ) result would be.

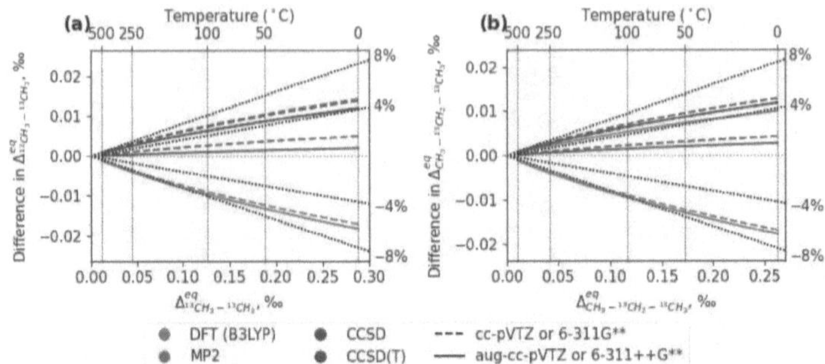


Figure 6.7: **Differences in carbon-13 clumped effect in ethane (a) and propane (b)** computed with the four electronic structure methods investigated in this study. All computations are harmonic relative to the Harmonic F12/ATZ. Dotted vertical lines label temperatures (from 0°C to 500°C). The slanted black lines denote relative difference in fractionation (4% and 8%) and the horizontal line is placed at y=0 where the reference (F12/ATZ) result would be.

### Harmonic frequency scaling factors

Figs. 6.8 and 6.9 address the effects of scaling the harmonic frequencies on the fractionation between alkanes (Fig. 6.8) and site-specific isotope effect (Fig. 6.9) of D (a-b) and $^{13}$C (c-d). In general, there is a small decrease in percent differences relative to the reference (F12/ATZ) method upon scaling (i.e., comparing the left panels to the right panels). The effect of scaling is most noticeable for the (least accurate) RHF method, bringing the relative error of this method close to that of the more accurate electronic structure methods.

Fig. 6.10 addresses the effects of scaling the harmonic frequencies on the clumped D + D (upper panels) and $^{13}$C + D (middle panels) and $^{13}$C + $^{13}$C (lower panels) heavy isotope effect in alkanes. The right-hand side panels utilize frequencies scaled by a constant (typically slightly smaller than 1) factor that depends on the electronic structure method and the basis set used, as tabulated in (REF). In general, scaling of the frequencies has a small impact on the relative differences for higher quality methods with larger basis sets and substantially improves agreement with the reference method for the lower quality methods with smaller basis sets. Comparing the left (unscaled frequencies) and right (scaled frequencies) panels, the RHF results benefit significantly from using the scaling factors. However, large deviations in D + D clumping using DFT with smaller basis sets persists after applying the scaling factors.

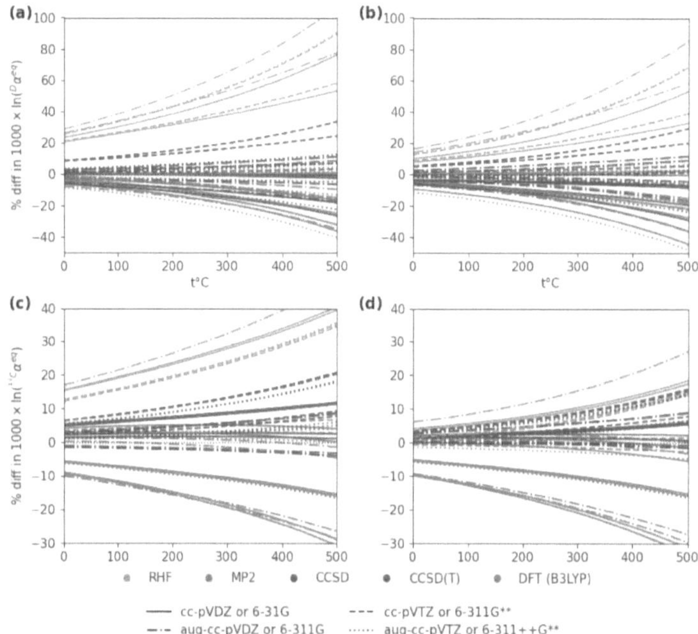

Figure 6.8: **Effect of scaling the harmonic frequencies on relative differences (in %) of fractionation of** D **(a-b), and** $^{13}$C **(c-d)** between ethane and methane as well as between propane and methane computed with different methods relative to the F12/ATZ method. Panels (a,c) are computed without scaling the frequencies, while panels (b,d) are based on the scaled frequencies.

### Differences due to anharmonic and non-Born-Oppenheimer effects

The diagonal Born-Oppenheimer correction increases the fractionation of deuterium between propane and methane ($^{D}\alpha^{eq}_{C_3H_8-CH_4}$) as well as between ethane and methane ($^{D}\alpha^{eq}_{C_2H_6-CH_4}$) by about 5% (Fig. 6.11a). Deuterium fractionations involving water are most substantially affected by the DBO correction, as exemplified by methane-water fractionation on Fig. 6.11(a). The value of $^{D}\alpha^{eq}_{CH_4-H_2O}$ is decreased by 10-30‰ due to the DBO correction (orange squares) over the range of temperatures studied. In contrast, the DBO correction does not change the fractionation of carbon-13 between alkanes (Fig. 6.11b).

The anharmonic effects decrease the fractionation of carbon-13 with methane ($^{13}\alpha^{eq}_{C_3H_8-CH_4}$ and $^{13}\alpha^{eq}_{C_2H_6-CH_4}$) by about 5% (Fig. 6.11b). Deuterium fractionation between methane and propane ($^{D}\alpha^{eq}_{C_3H_8-CH_4}$) is affected significantly (by about

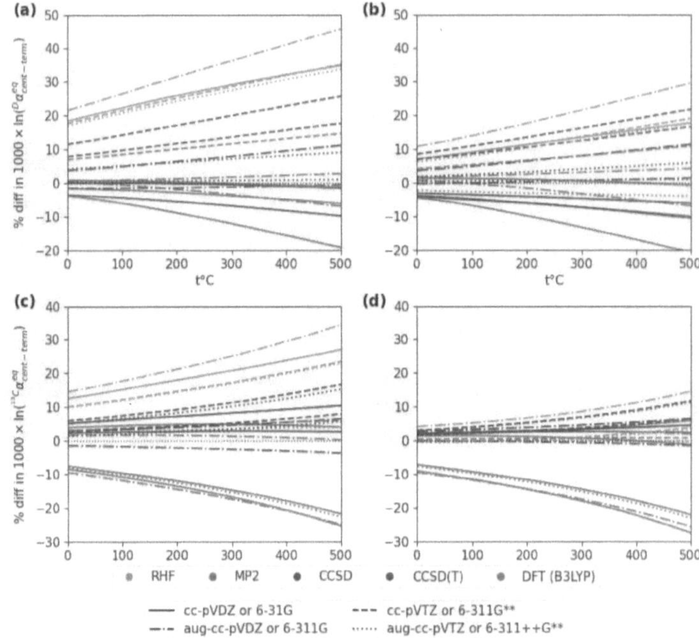



Figure 6.9: **Effect of scaling the harmonic frequencies on relative differences (in %) of site-specific isotope effect of** D **(a-b), and** $^{13}$C **(c-d)** in propane computed with different methods relative to the F12/ATZ method. Panels (a,c) are computed without scaling the frequencies, while panels (b,d) are based on the scaled frequencies.

20%), while other fractionations ($^D\alpha^{eq}_{H_2O-CH_4}$ and $^D\alpha^{eq}_{C_2H_6} - CH_4$) are affected much less (up to -5% relative difference), see Fig. 6.11a. Note that for deuterium fractionations between alkanes the DBO and anharmonic effects have the opposite sign and therefore they partially cancel out.

Position-specific isotope effect in propane is affected by both the DBO correction and the anharmonic effects and the effects happen to align, reinforcing each other relative to the harmonic result. For carbon-13 the DBO correction decreases $^{13}\alpha^{eq}_{center-terminal}$ by a little over 1% in relative or 0.1-0.2‰ in the absolute sense and the anharmonic effects are responsible for another 3% relative (0.2-0.4‰ absolute) decrease (see Fig. 6.12b). For D position-specific effect the effect is even more pronounced with both DBO correction and the anharmonic effect increasing $^D\alpha^{eq}_{center-terminal}$ by about 5% (or 5-10‰) each (Fig 6.12a).

We verify that the clumped heavy isotope effects are not affected by the DBO cor-

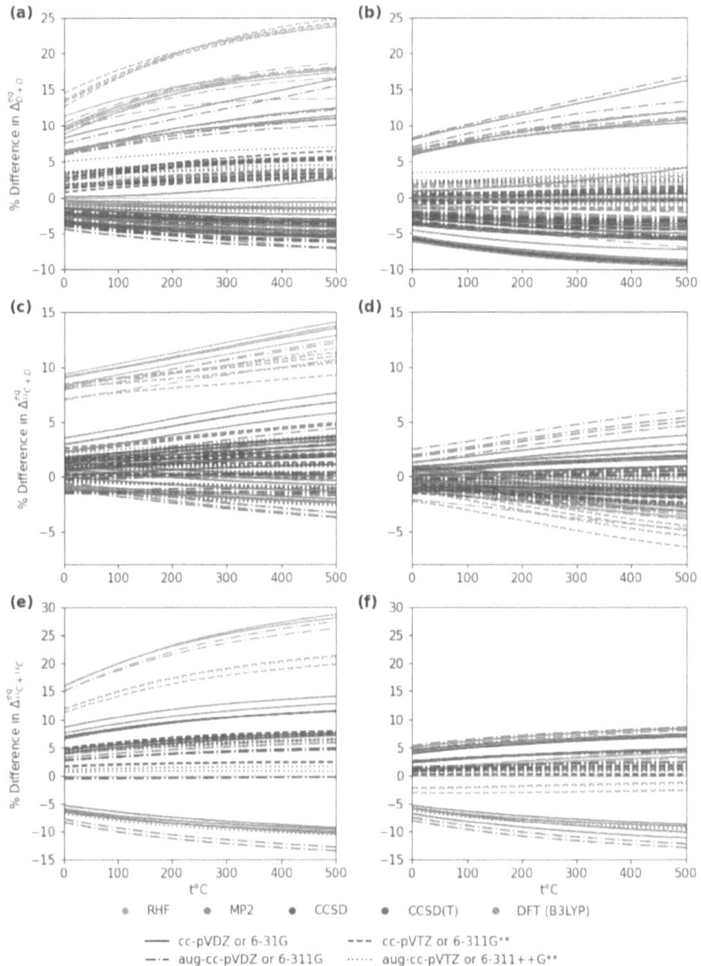

Figure 6.10: **Effect of scaling the harmonic frequencies on relative differences (in %) of clumped** $D+D$ **(a-b),** $^{13}C+D$ **(c-d) and** $^{13}C+^{13}C$ **(e-f) isotope effects in alkanes** computed with different methods relative to the F12/ATZ method. Panels (a,c,e) are computed without scaling the frequencies, while panels (b,d,f) are based on the scaled frequencies.

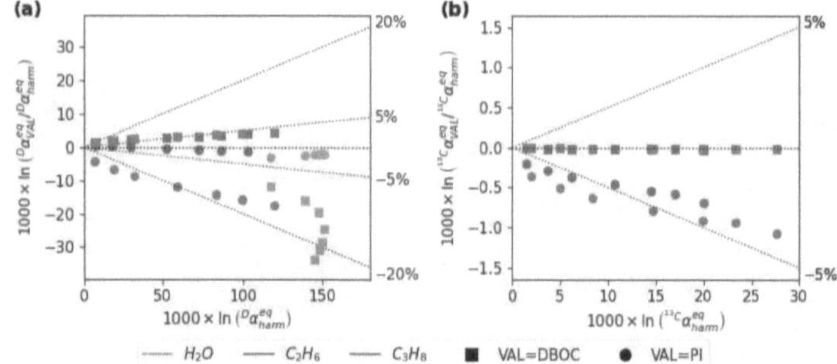

Figure 6.11: **Effect of the diagonal Born-Oppenheimer correction (DBOC, squares) and anharmonic contributions (PI, circles) on the fractionation of deuterium (a) and (b) carbon-13** with methane for water ($\alpha^{eq}_{H_2O-CH_4}$), ethane ($\alpha^{eq}_{C_2H_6-CH_4}$) and propane ($\alpha^{eq}_{C_3H_8-CH_4}$). The following temperatures are shown: 500°C, 300°C, 200°C, 100°C, 50°C, 25°C, 0°C (left to right). The slanted black lines denote relative difference in fractionation (5 and 20%) and the horizontal line is placed at y=0 where the reference (harmonic F12/ATZ) result would be.

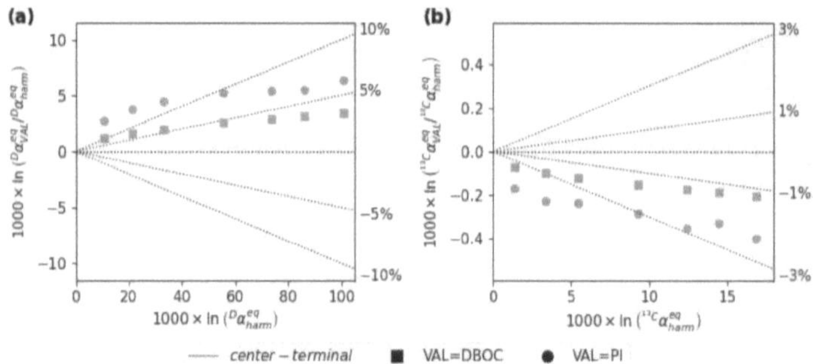

Figure 6.12: **Effect of the diagonal Born-Oppenheimer correction (DBOC, squares) and anharmonic contributions (PI, circles) on the position-specific isotope effect of deuterium (a) and carbon-13 (b) in propane.** The following temperatures are shown: 500°C, 300°C, 200°C, 100°C, 50°C, 25°C, 0°C (left to right). The slanted black lines denote relative difference in fractionation (1%, 3%, 5% and 10%) and the horizontal line is placed at y=0 where the reference (harmonic F12/ATZ) result would be.

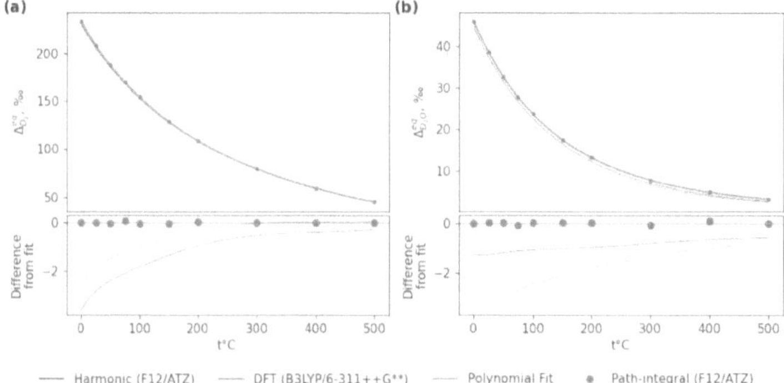


Figure 6.13: **Fifth order least squares polynomial fit (in gray) to the clumped** $D + D$ **isotope effect in dihydrogen (a) and water (b) calculated with PIMC (red circles).** The harmonic result with the same electronic structure method is shown in blue and the best DFT result obtained in this study in green. The bottom panels show the same data, but the polynomial fit is subtracted off.

rection.[179] Anharmonic effects are most pronounced in $D + D$ clumping of hydrogen and water (Fig. 6.13), i.e., up to 4‰ and up to 2‰ at the lowest temperature (0°C), respectively. Full anharmonic treatment of all other clumped isotope effects is comparable to the harmonic result within statistical uncertainty of the path-integral calculations (see Figs. 6.14 and B.12, B.13, B.14, B.15). It is possible to reduce the size of statistical uncertainties by sampling further, but we do not attempt to do so here and instead just conclude that the anharmonic effects contribute less than 0.1‰ for $^{17}O + D$ clumping in water (Fig. B.12a), $^{13}C + D$ clumping in methane (Fig. 6.14a), less than 0.3‰ for $^{18}O + D$ clumping in water (Fig. B.12b), $^{13}C + D$ clumping in ethane (Fig. 6.14b) and $D + D$ clumping in methane (Fig. B.13a). Finally, the bounds on $D + D$ clumping in ethane (Fig. B.13b) and propane (Fig. B.14) as well as $^{13}C + D$ clumping in propane (Fig. B.15) are about $\pm 1$‰.

We also note that all other clumped isotope effects in ethane and propane (i.e., all $^{13}C + {}^{13}C$ effects as well as $D + D$ effects where the two deuterium atoms are not bound to the same carbon and $^{13}C + D$ effects where the heavy atoms are not directly bound to each other) have small magnitude ($<0.5$‰). The small size of the effect (similarly to a weak signal) exacerbates the relative importance of statistical uncertainty even further, making the path-integral results unreliable. We return to discuss this in Section 6.5.

**Best results**

Tables 6.6 and 6.7 provide our best estimates for the fractionation values between alkanes for deuterium and carbon-13, respectively. These values are obtained from

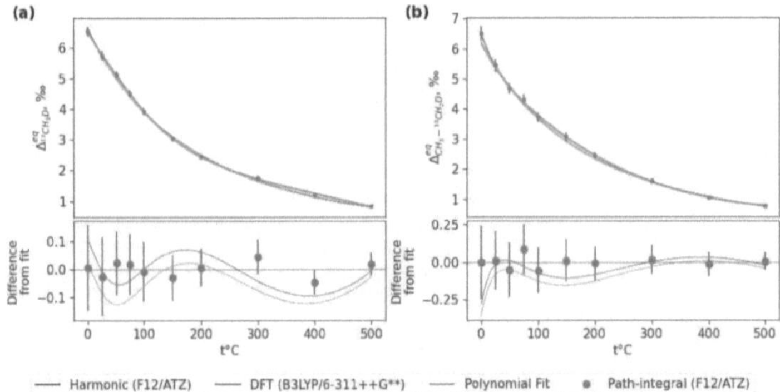

Figure 6.14: **Fifth order least squares polynomial fit (in gray) to the clumped $^{13}$C + D isotope effect in methane (a) and ethane (b) calculated with PIMC (red circles).** The harmonic result with the same electronic structure method is shown in blue and the best DFT result obtained in this study in green. The bottom panels show the same data, but the polynomial fit is subtracted off.

| t,°C | $1000 \times \ln{}^D\alpha^{eq}_{C_2H_6-CH_4}$ | | | $1000 \times \ln{}^D\alpha^{eq}_{C_3H_8-CH_4}$ | | | $1000 \times \ln{}^D\alpha^{eq}_{C_3H_8-C_2H_6}$ | | |
|---|---|---|---|---|---|---|---|---|---|
| 0 | 105.5 | ± | 1.34E-01 | 106.67 | ± | 1.62E-01 | 1.17 | ± | 1.92E-01 |
| 25 | 88.61 | ± | 1.37E-01 | 87.63 | ± | 1.58E-01 | -0.99 | ± | 1.80E-01 |
| 50 | 75.06 | ± | 1.16E-01 | 72.71 | ± | 1.33E-01 | -2.35 | ± | 1.56E-01 |
| 75 | 63.93 | ± | 1.03E-01 | 60.41 | ± | 1.10E-01 | -3.52 | ± | 1.29E-01 |
| 100 | 54.98 | ± | 1.03E-01 | 50.68 | ± | 1.08E-01 | -4.3 | ± | 1.16E-01 |
| 150 | 41.53 | ± | 9.17E-02 | 36.135 | ± | 8.92E-02 | -5.395 | ± | 9.99E-02 |
| 200 | 32.081 | ± | 7.21E-02 | 26.116 | ± | 7.53E-02 | -5.964 | ± | 8.42E-02 |
| 300 | 20.16 | ± | 6.23E-02 | 14.126 | ± | 6.09E-02 | -6.034 | ± | 6.60E-02 |
| 400 | 13.474 | ± | 5.18E-02 | 7.786 | ± | 4.99E-02 | -5.688 | ± | 5.64E-02 |
| 500 | 9.355 | ± | 3.99E-02 | 4.176 | ± | 3.96E-02 | -5.179 | ± | 4.47E-02 |

Table 6.6: Deuterium fractionation between alkanes calculated at the F12/ATZ level of theory with PIMC and the DBO correction.

the path-integral calculations at the F12/ATZ level of theory with the DBO correction. The individual sites of propane are calculated separately and averaged as described in Appendix A.2. The largest alkane (propane) is the heaviest in terms of carbon-13 fractionation, followed by ethane, which makes methane the lightest (Table 6.7). On the other hand, for deuterium fractionation propane is lighter than ethane over the temperature range considered here (see Table 6.6). Table 3 provides the values for site-specific isotope effect in propane. The center site of propane is significantly heavier than the terminal both for deuterium and carbon-13 site-specific isotope effects.

| $t,^\circ C$ | $1000 \times \ln {}^{13}\alpha^{eq}_{C_2H_6-CH_4}$ | | $1000 \times \ln {}^{13}\alpha^{eq}_{C_3H_8-CH_4}$ | | $1000 \times \ln {}^{13}\alpha^{eq}_{C_3H_8-C_2H_6}$ | |
|---|---|---|---|---|---|---|
| 0 | 19.323 | ± 3.14E-02 | 26.503 | ± 2.91E-02 | 7.18 | ± 2.76E-02 |
| 25 | 16.396 | ± 2.69E-02 | 22.397 | ± 2.50E-02 | 6.001 | ± 2.68E-02 |
| 50 | 13.946 | ± 2.49E-02 | 18.962 | ± 2.36E-02 | 5.016 | ± 2.42E-02 |
| 75 | 11.907 | ± 2.07E-02 | 16.16 | ± 2.19E-02 | 4.252 | ± 2.11E-02 |
| 100 | 10.264 | ± 1.92E-02 | 13.86 | ± 1.83E-02 | 3.596 | ± 1.83E-02 |
| 150 | 7.641 | ± 1.56E-02 | 10.288 | ± 1.44E-02 | 2.646 | ± 1.45E-02 |
| 200 | 5.771 | ± 1.15E-02 | 7.724 | ± 1.11E-02 | 1.952 | ± 1.26E-02 |
| 300 | 3.402 | ± 1.00E-02 | 4.4966 | ± 9.28E-03 | 1.0947 | ± 8.39E-03 |
| 400 | 2.052 | ± 7.25E-03 | 2.6624 | ± 7.12E-03 | 0.6105 | ± 6.19E-03 |
| 500 | 1.2558 | ± 8.13E-03 | 1.6087 | ± 7.34E-03 | 0.3529 | ± 6.18E-03 |

Table 6.7: Carbon-13 fractionation between alkanes calculated at the F12/ATZ level of theory with PIMC and the DBO correction.

| $t,^\circ C$ | $1000 \times \ln {}^{D}\alpha^{eq}_{center-terminal}$ | | $1000 \times \ln {}^{13}\alpha^{eq}_{center-terminal}$ | |
|---|---|---|---|---|
| 0 | 110.47 | ± 3.39E-01 | 16.298 | ± 3.90E-02 |
| 25 | 94.58 | ± 3.29E-01 | 13.946 | ± 3.43E-02 |
| 50 | 82.09 | ± 2.51E-01 | 11.917 | ± 3.30E-02 |
| 75 | 71.74 | ± 2.28E-01 | 10.318 | ± 3.25E-02 |
| 100 | 63.24 | ± 1.82E-01 | 8.895 | ± 2.41E-02 |
| 150 | 49.31 | ± 1.52E-01 | 6.692 | ± 1.99E-02 |
| 200 | 39.88 | ± 1.42E-01 | 5.113 | ± 1.76E-02 |
| 300 | 26.96 | ± 1.07E-01 | 3.0179 | ± 9.63E-03 |
| 400 | 19.102 | ± 9.31E-02 | 1.8366 | ± 8.62E-03 |
| 500 | 14.391 | ± 7.76E-02 | 1.1242 | ± 6.72E-03 |

Table 6.8: Position-specific isotope effect in propane calculated at the F12/ATZ level of theory with PIMC and the DBO correction.

We also provide the $5^{th}$ order interpolative polynomial fit coefficients for $1000 \times \ln RPFR(A^*)$ needed to calculate carbon-13 and deuterium fractionation as well as the site-specific isotope effect in propane (Tables B.1 and B.2). The best-fit polynomials produce fits with largest residuals of <0.03‰. There are two advantages of fitting individual molecules and not pairs of molecules: (i) fewer fits; (ii) new molecules can be easily added later. The fractionation for any pair can be obtained by first subtracting the coefficients of the "bottom" molecule from the "top" at each polynomial order and then plugging in the temperature into resulting $5^{th}$ order polynomial equation.

For the clumped heavy isotope effects we only provide the polynomial fits for D + D clumping in dihydrogen and water in Table 6.9; the PIMC-based values are in Table 6.10. In all other cases we recommend using the BMU (harmonic) formula (Eq. 3.25) based on the frequencies obtained via F12/ATZ method.

| Poly order | $0^{th}$ | $1^{st}$ | $2^{nd}$ | $3^{rd}$ | $4^{th}$ | $5^{th}$ | Max Res |
|---|---|---|---|---|---|---|---|
| $H_2$ | -22.917088 | 19.324750 | 37.531448 | -10.852238 | 1.4122405 | -0.0588901 | 0.067 |
| $H_2O$ | 11.861599 | -26.307850 | 21.333101 | -6.0087275 | 1.120047 | -0.0942403 | 0.084 |

Table 6.9: Fifth order least squares fit coefficients for the D + D clumped isotope effect in molecular hydrogen and water. Last column contains the value of the maximum residual from fit.

| $t, °C$ | $^{D}\Delta^{eq}_{D_2}$ | | $^{D}\Delta^{eq}_{D_2O}$ | |
|---|---|---|---|---|
| 0 | 233.328 | $\pm$ 9.77E-02 | 45.86 | $\pm$ 1.43E-01 |
| 25 | 208.361 | $\pm$ 7.72E-02 | 38.65 | $\pm$ 1.08E-01 |
| 50 | 187.47 | $\pm$ 6.17E-02 | 32.67 | $\pm$ 1.08E-01 |
| 75 | 169.744 | $\pm$ 5.16E-02 | 27.651 | $\pm$ 9.61E-02 |
| 100 | 154.209 | $\pm$ 4.91E-02 | 23.687 | $\pm$ 8.43E-02 |
| 150 | 128.809 | $\pm$ 3.60E-02 | 17.549 | $\pm$ 8.15E-02 |
| 200 | 108.845 | $\pm$ 3.63E-02 | 13.216 | $\pm$ 6.90E-02 |
| 300 | 79.549 | $\pm$ 2.55E-02 | 7.759 | $\pm$ 6.41E-02 |
| 400 | 59.479 | $\pm$ 2.11E-02 | 5.014 | $\pm$ 4.85E-02 |
| 500 | 45.128 | $\pm$ 1.83E-02 | 3.288 | $\pm$ 4.07E-02 |

Table 6.10: D + D clumped isotope effect in molecular hydrogen and water.

## 6.4 Discussion

When calculating the RPFR's for real molecules, we are limited to approximate treatments of the problem. Therefore, it is a focus of this study to assess the relative importance of the approximations that are frequently made when reporting calculated RPFR's in the literature.

**Approximate description of the molecular potential**

We begin with discussing the approximations of potential energy of the molecule as a function of its molecular geometry. Theoretical chemistry offers a variety of methods to calculate the potential energy, depending on the accuracy requirements and the availability of computational resources. Figs. 6.2-6.7 in the main text and B.2-B.11 in the Appendix address the effect of using select theoretical methods to describe the molecular potential. We presented the difference in values of $1000 \times \ln \alpha^{eq}$ and $\Delta^{eq}$ computed with different levels of theory and basis set sizes relative to the reference (F12/ATZ) method in those figures. There are two caveats with such analysis: (i) we do not know the true values, therefore we present comparisons relative to the F12/ATZ result and (ii) this comparison only identifies differences in the fractionation factors computed based on the harmonic approximation, i.e., those that arise due to the different curvature of the potentials at the minimal energy (equilibrium) geometry. Any anharmonic effects of the potential are not accounted for in this comparison. With that caveat in mind, a few important conclusions are listed below.

Generally, isotopic fractionations and position-specific isotope effects computed with DFT (B3LYP functional) or the post-Hartree-Fock methods (MP2, CCSD, CCSD(T)) are within 10% relative difference, while clumped isotope effects are typically within 5% from the reference method (F12/ATZ). The basis set size does not matter significantly, other than for the DFT (B3LYP) calculation, where smaller (6-31G and 6-311G) basis sets yield significant deviations from the reference values. DFT/B3LYP is the method with the smallest computational cost out of the methods whose performance is generally adequate. However, the equilibria involving Carbon-13 (bulk fractionation, site-specific and $^{13}C + D$ as well as $^{13}C + {}^{13}C$ clumping) tend to be underestimated by about 10% relative to the reference result. We do not recommend decreasing the quality of the molecular potential to empirical force field (AIREBO, CHARMM) or the mean-field (RHF) methods. These can yield differences in isotopic equilibria from those predicted by the higher quality methods, that overwhelm all other approximations. There is no uniform trend with less accurate methods being farther away and more accurate methods clustering around the unknown "true answer." In particular, the MP2 results are typically closer to the reference F12 values than those calculated using the CCSD method, which is typically considered to be more accurate. This highlights the unpredictable nature of error cancellations, which can yield the "right" result for the "wrong" reasons.

**Cancellation of errors in molecular potentials**

All the results presented in Figs. 6.2-6.7 and B.2-B.11 are based on the harmonic frequencies, which are very sensitive to the correct description of the molecular potential. Indeed, we observe large differences in harmonic frequencies of the molecules computed with different methods.This suggests that proper description of the molecular potential is very important in getting the isotopic equilibria right. On the other hand, errors in describing molecular potential have a similar effect on all the isotopologues involved in the equilibrium reaction, such that even though the harmonic frequencies from two different methods differ substantially, there is error cancellation when computing the RPFR's and a second time when calculating their ratios. The error cancellation can lead to excellent agreement between the harmonic approximation and the fully quantum mechanical description of clumping in methane as shown by [86] for the $^{13}C + D$ and by [98] for the $D + D$ clumped effects in methane. Figure 6.15 addresses the question of the importance of error cancellation when calculating the equilibrium fractionation factors $1000 \times \ln \alpha^{eq}$ in alkanes. We plot the RPFR's for each singly substituted isotopologue relative to the unsubstituted one. It is the ratio of these RPFR's that determine the fractionation factors (see Eq. 3.21). The first thing to note is that the absolute differences in RPFRs of alkanes are much larger (up to 60‰ for deuterium and 3‰ for carbon-13) than the differences in fractionation factors. The latter are only up to 10‰ for deuterium and 1.5‰ for carbon-13 (see Fig. 6.2). This means that most (over half) of the difference in RPFR values cancel out when fractionation factors are computed

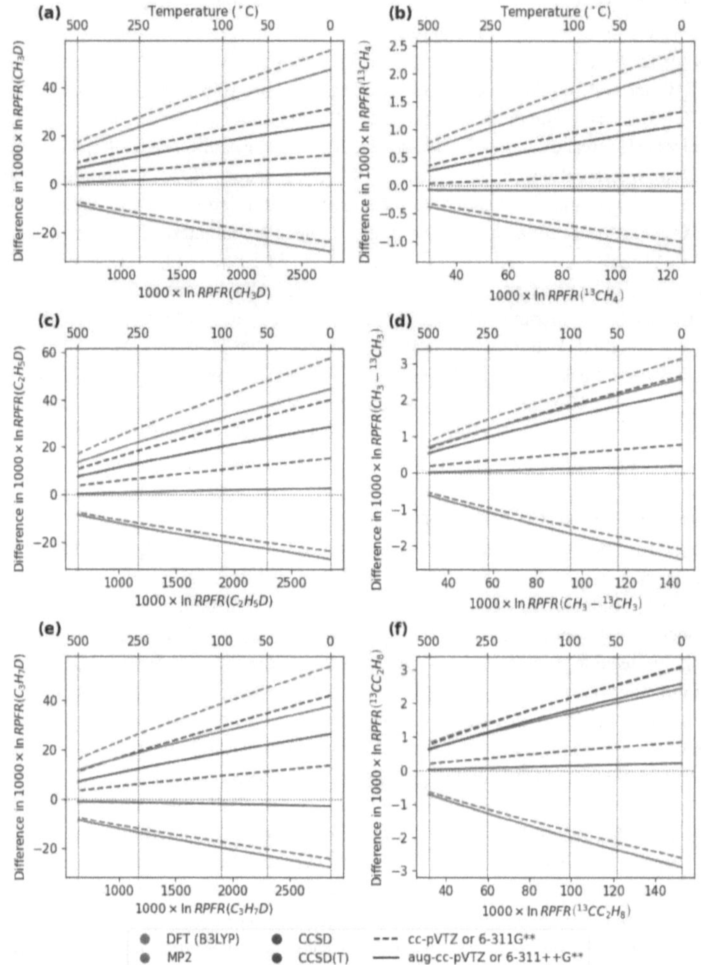

Figure 6.15: **Difference in harmonic** $1000 \times \ln(\text{RPFR})$ **of methane (a-b), ethane (c-d) and propane (e-f)** with a single heavy atom (deuterium in a,c,e and carbon-13 in b,d,f) computed with 4 commonly used electronic structure methods relative to the F12/ATZ method. Dotted vertical lines label temperatures (from 0°C to 500°C).

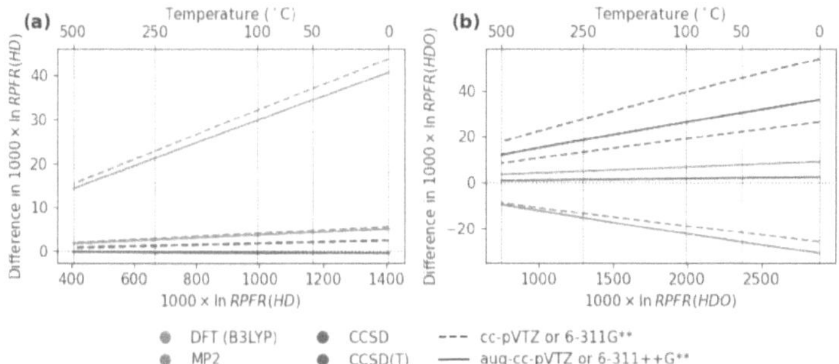


Figure 6.16: **Difference in harmonic** $1000 \times \ln(\text{RPFR})$ **of monodeuterated molecular hydrogen (a) and water (b)** computed with 4 commonly used electronic structure methods relative to the F12/ATZ method. Dotted vertical lines label temperatures (from 0°C to 500°C).

using Eq. (3.21), provided the same method is used to describe both molecules. This also explains why we do not observe consistently better agreement with the reference method when the basis set size is increased (Fig. B.2). Even though the difference in RPFR relative to the reference is consistently larger for the cc-pVTZ than for the aug-cc-pVTZ (Fig. 6.15), we do not see the same trends on Fig. 6.2. Including diffusive functions (i.e., going from 6-311G$^{**}$ to 6-311++G$^{**}$) in the DFT calculations does not substantially change the RPFR's of alkanes, but in all cases the change is farther away from the reference.

The large error cancellation when converting RPFR's to the fractionation values practically means that a lower-cost calculation can be used with none to minimal loss of accuracy. On the flip side, the differences in fractionation values of alkanes (Figs. 6.2, B.2 and B.1) are less predictable. The RPFR's of alkanes with one heavy isotope (either D or $^{13}$C) converge from above to the F12/ATZ value both with improving the quality of the electronic structure method from RHF (not shown on Fig. 6.15 due to large error) to MP2 to CCSD to CCSD(T) and with increasing the basis set size from cc-pVTZ to aug-cc-pVTZ. DFT calculations do not fit into the hierarchy of the post-Hartree-Fock methods, and they underestimate the RPFR's relative to the reference value by the magnitude comparable to that of the CCSD method. Because the extent to which the errors cancel does not improve with increasing the level of theory or basis set size the trends in RPFR values are obscured by the large effects of error cancellation. It is this error cancellation that makes it possible for a less accurate method to get the "right answer" for the "wrong reasons," so one must be very careful not to assume that if the method (say MP2) was sufficiently accurate in one case, it ought to be similarly accurate in others.

This is exemplified by the alkane-water fractionation, which varies dramatically across the methods (see Figs. 6.3, B.4 and B.3). Because alkanes are much more alike, it is expected that they would manifest especially favorable error cancellation, while water has different molecular properties (including a strong dipole moment). As a result, the RPFR values of water show a different error pattern (see Fig. 6.16(b)) and the error cancellation for fractionations between water and other molecules is not as strong. This is observed in particular for MP2 which yields fractionations involving water up to 40‰ (in absolute terms) or 25% (in relative terms) different from the reference, standing out from the other theoretical models (see Fig. 6.3).

Even though the relative differences in fractionation factors that involve dihydrogen are small due to the large magnitude of the equilibrium fractionation factors, the absolute differences in $1000 \times \ln {}^{D}\alpha^{eq}_{\text{molecule}-\text{H}_2}$ (Fig. 6.3a) are as large as the differences in corresponding RPFRs (Figs. 6.15 and 6.16a) indicating poor error cancellation. This is an issue unique to molecular hydrogen and it arises because $H_2$ only has two electrons, so the correlation between them is described much better by all post-Hartree-Fock methods. Moreover, CCSD provides an exact description of correlation (since 3 or more electrons cannot be excited in a system that only has 2 electrons) at a given basis set size.

The clumped heavy isotope effect yields even better error cancellation, generally yielding differences between molecular potentials of less than 5%. We believe this holds for other molecules beyond those considered here. However, the site-specific effect in propane (Fig. 6.4) is as sensitive to the potential as the fractionations between different alkanes (Fig. 6.2). Figure 6.4 shows that there is substantial, but not excellent error cancellation when the ratio of RPFR is computed, just as was the case with the fractionation of alkanes.

Fig. 6.17 summarizes these trends with relative error for fractionations (a-b), position-specific effect (c-d) and clumping (e-f) in alkanes involving deuterium (left panels) and carbon-13 (right panels). The relative errors span a range of 30-40% for the fractionation, while clumped isotope effect has a significantly narrower range of $\pm 10\%$ for D + D clumping (panel e) and $\pm 5\%$ for $^{13}C + D$ clumping (panel f). The distribution of relative errors for the site-specific isotope effect (Fig. 6.17c-d) is more similar to the fractionation (panels a-b) than to clumping (panels e-f). This is especially true for carbon-13, where the distributions of errors on panels (b) and (d) are almost identical, while panel (f) has a narrow distribution without the second peak at negative relative errors.

Fig. 6.18 addresses whether the size of the relative difference in fractionation (irrespective of the sign) is correlated with the size of the relative difference in clumping (panels a-b) as well as the site-specific isotope effect (panels c-d). In general, the differences in clumped effects do not correlate well with the differences in corresponding fractionations. This is especially pronounced for the MP2 method (green) as the relative importance of the error cancellation is the largest for this method. In contract, differences site specific affects correlate well with the differences in

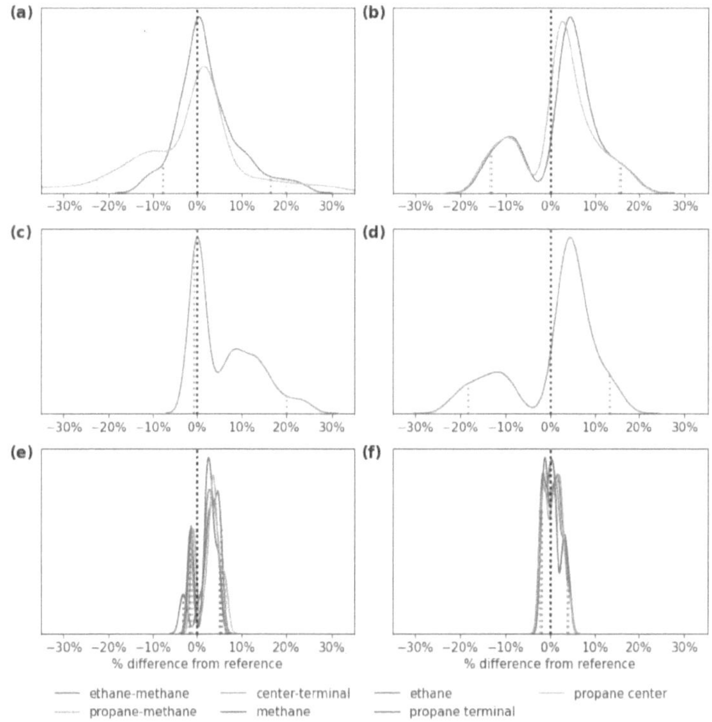


Figure 6.17: **Relative deviations from the reference value (dotted black line) for fractionations (a-b), position-specific (c-d) and clumped (e-f) isotope effects in alkanes** involving deuterium (left panels) and carbon-13 (right panels). Harmonic calculations over the temperature range of 0-500°C are done using DFT (B3LYP) with 6-311G** and 6-311++G** basis sets as well as MP2, CCSD and CCSD(T) with cc-pVTZ and aug-cc-pVTZ basis sets each. Dotted lines are 5th and 95th percentile of the data on each distribution.

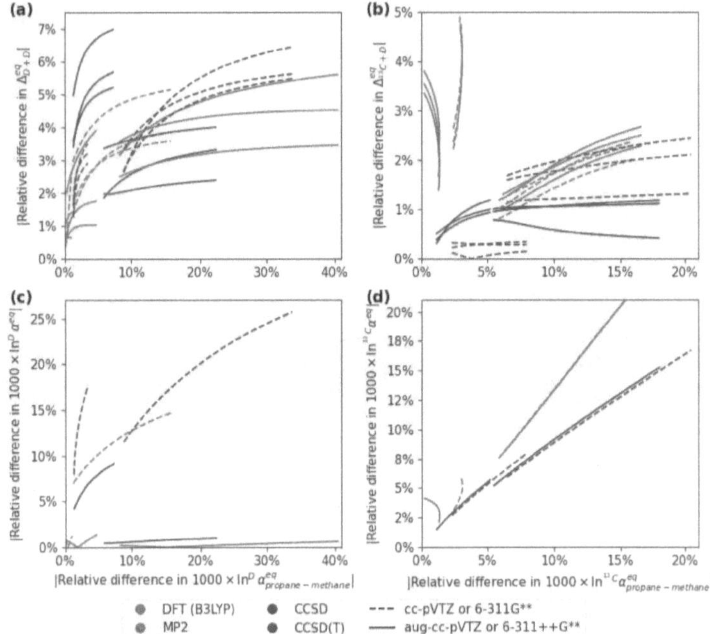

Figure 6.18: **Absolute values of the relative (%) differences in clumped (a-b) and site-specific (c-d) isotope effects** plotted against the absolute values of the relative differences in fractionation values between propane and methane over the temperature range of 0 to 500°C. All differences are relative to the F12/ATZ reference method. Panel (a) contains the clumped D+D isotope effect in methane as well as center and terminal positions of propane. Panel (b) contains the analogous $^{13}$C + D clumped effects.

fractionation, especially for the carbon-13 equilibria (panel d). Here all methods except for MP2 show strong correlation between $1000 \times \ln {}^{13}\alpha^{eq}_{\text{propane-ethane}}$ and $1000 \times \ln^{13} \alpha^{eq}_{\text{center-terminal}}$.

## Scaling of the harmonic frequencies

We used the most common and simplest procedure for scaling the harmonic frequencies, wherein all frequencies are scaled by the same factor that depends on the electronic structure method and the basis set used. These factors are computed by minimizing the difference in harmonic frequencies relative to a reference for a number of molecules. As discussed in the previous section, methods that get closer to the true harmonic frequencies of the molecule do not necessarily yield more accurate isotopic equilibria due to significant role of error cancellation. Thus,

| Method | Clumping | | | | Fractionation | | | | Site specific effect | | | |
|---|---|---|---|---|---|---|---|---|---|---|---|---|
| | 0°C | | 500°C | | 0°C | | 500°C | | 0°C | | 500°C | |
| | pre | post | pre | post | pre | post | pre | post | pre | post | pre | post |
| RHF | 9.55 | 1.52 | 14.84 | 3.63 | 16.01 | 6.72 | 35.52 | 25.33 | 13.11 | 4.39 | 24.1 | 11.78 |
| B3LYP | 3.2 | 2.95 | 5.94 | 5.56 | 4.81 | 4.94 | 22.82 | 24.8 | 5.31 | 5.52 | 19.79 | 20.66 |
| MP2 | 2.8 | 1.77 | 5.11 | 3.27 | 3.25 | 3.52 | 16.65 | 21.46 | 2.45 | 2.1 | 4.55 | 4.17 |
| CCSD | 2.39 | 1.74 | 3.91 | 2.89 | 3.99 | 2.84 | 11.64 | 10.37 | 4.57 | 3.4 | 10.42 | 8.9 |
| CCSD(T) | 1.68 | 1.36 | 2.79 | 2.38 | 2.79 | 2.26 | 10.48 | 10.21 | 2.58 | 2.06 | 5.83 | 5.49 |

Table 6.11: Mean absolute values of relative differences (%) between isotopic equilibria in alkanes pre- and post- the application of harmonic frequency scaling factors at lowest and highest temperatures studied. Note that the frequencies computed by the reference method (F12/ATZ) also get scaled to yield the "post" column. Data over all four basis sets for each electronic structure method studied were averaged to yield 40 data points for the clumped isotope effects (D + D, $^{13}$C + D and $^{13}$C + $^{13}$C), 16 fractionation factors (propane-methane and ethane-methane for both deuterium and carbon-13) and 8 site-specific effects in propane (carbon 13 and deuterium).

it is not obvious that equilibria computed with scaled frequencies are necessarily closer to the true values. Figures 6.9-6.10 show that scaling has the biggest effect on the results from RHF, i.e., poorest quality electronic structure method studied here. In practical terms this is not particularly useful as the DFT-based methods typically provide superior results at comparable cost to the RHF method. This is why harmonic frequency calculations are almost never done with the RHF method. Interestingly, the scaling does not significantly reduce the difference between the DFT (B3LYP) and all other methods for isotopic equilibria that involve carbon-13 (bottom two panels of all three figures). The effect on all other methods is small, but generally positive, in a sense that the difference from the reference result generally decreases upon scaling of both. It is also not surprising that the results computed with smaller basis sets (solid and dash-dotted lines) typically yield more significant improvements towards agreement with the (post-scaling) reference method. Table 6.11 highlights the trends from Figures 3.7-3.9. While agreement with the reference method is significantly improved for the RHF method, the DFT (B3LYP) is not significantly affected by the scaling of harmonic frequencies. Overall, it appears that clumped isotope effect largely benefits from scaling the harmonic frequencies of alkanes in a sense that equilibria computed with the scaled frequencies are typically closer to the (also scaled) reference value. At the same time, the impact of the scaling factors decreases with increasing quality of the computational method used. The situation is much less clear for the fractionation factors and the site-specific isotope effect, where frequency scaling can either improve the agreement (e.g., for RHF) or worsen it (e.g., for MP2).

While it is possible to use more advanced scaling schemes (e.g., including frequency and/or molecule dependence), we do not analyze their effects in this study. The scaling factors we used are provided in Table 6.12. Frequency scaling can be a practical solution to achieving better apparent agreement between different methods,

| | 6-31G | 6-311G | 6-311G(d,p) | 6-311++G(d,p) |
|---|---|---|---|---|
| B3LYP | 0.962 | 0.966 | 0.967 | 0.967 |
| | cc-pVDZ | cc-pVTZ | aug-cc-pVDZ | aug-cc-pVTZ |
| MP2 | 0.95 | 0.949 | 0.969 | 0.951 |
| CCSD | 0.947 | 0.948 | 0.963 | 0.951 |
| CCSD(T) | 0.963 | 0.958 | 0.971 | 0.964 |
| F12 | | | | 0.9638 |

Table 6.12: Harmonic frequency scaling factors used in this study and obtained from Ref. [210].

but it is a crude approach, that is only applicable to the harmonic approximation and hard to combine with approaches that go beyond it. Note that frequency scaling changes the reference result significantly. For example, D + D clumped isotope effect of methane at F12/ATZ (reference) method and a temperature of 0°C changes from 23.7‰ prior to scaling to 21.9‰ upon scaling of the harmonic frequencies. The shift in magnitude of values of isotopic equilibria is less important than the temperature dependence, which changes less dramatically upon scaling. Therefore, it is crucial to stay within either only scaled or only unscaled frequencies and to not mix data from both. We note that except for this section and the corresponding section in the results, everything presented in this book is based on the unscaled (true) harmonic frequencies.

**Importance of the DBO correction**

We observe results that are largely consistent with our intuition on the importance of the DBO correction described in Chapter 5. Recall that it is most prominent for fractionation of deuterium and for the fractionations between molecules of different polarity (i.e., between water and all other molecules). Unsurprisingly, the DBO correction has a negligible effect on fractionation of carbon-13 (Fig. 6.11b). Perhaps more surprisingly, the site-specific carbon-13 isotope effect in propane is affected by a small but not negligible amount of about 0.1‰ over the range of temperatures studied, which corresponds to 1-2% relative error (Fig. 6.12). We also note that the influence of the DBO correction is in some cases less noticeable due to (at least partial) cancellation with the anharmonic effects discussed in the following section, while in other cases (notably site-specific isotope effect) the two effects reinforce each other.

Tables 6.13 and 6.14 address how each of the individual singly substituted RPFR's is affected by the DBO correction. The total effect on the $\alpha^{eq}$ value for a pair of molecules is therefore just a difference between corresponding columns in the table.

**Importance of the anharmonic effects**

The PIMC method takes into account the anharmonic effects, that are missing in the approximate BMU (harmonic) treatment. However, the isotopic equilibria

| t °C | Hydrogen | Water | Methane | Ethane | Center | Terminal | Total |
|---|---|---|---|---|---|---|---|
| 0 | -0.67 | -33.72 | 0 | 7.23 | 6.95 | 3.5 | 4.44 |
| 25 | -0.61 | -30.9 | 0 | 6.63 | 6.36 | 3.21 | 4.05 |
| 50 | -0.56 | -28.51 | 0 | 6.11 | 5.87 | 2.96 | 3.73 |
| 75 | -0.52 | -26.46 | 0 | 5.68 | 5.45 | 2.75 | 3.46 |
| 100 | -0.49 | -24.69 | 0 | 5.3 | 5.08 | 2.56 | 3.22 |
| 150 | -0.43 | -21.77 | 0 | 4.67 | 4.48 | 2.26 | 2.84 |
| 200 | -0.38 | -19.47 | 0 | 4.18 | 4.01 | 2.02 | 2.53 |
| 300 | -0.32 | -16.07 | 0 | 3.45 | 3.31 | 1.67 | 2.09 |
| 400 | -0.67 | -33.72 | 0 | 7.23 | 6.95 | 3.5 | 4.44 |
| 500 | -0.61 | -30.9 | 0 | 6.63 | 6.36 | 3.21 | 4.05 |

Table 6.13: Changes in $1000 \times \ln \mathrm{RPFR}(A^*)$ due to the DBO correction for single deuterium substitution. For convenience, the effect is given relative to methane (i.e., the effect on $1000 \times \ln \mathrm{RPFR}(\mathrm{CH_3D})$ is subtracted from each value. "Center," "Terminal," and "Total" refer to the two positions of propane and their appropriate average.

| t °C | Methane | Ethane | Center | Terminal | Total |
|---|---|---|---|---|---|
| 0 | 0 | -0.023 | 0.05 | -0.155 | -0.019 |
| 25 | 0 | -0.021 | 0.046 | -0.142 | -0.017 |
| 50 | 0 | -0.02 | 0.042 | -0.131 | -0.016 |
| 75 | 0 | -0.018 | 0.039 | -0.122 | -0.015 |
| 100 | 0 | -0.017 | 0.037 | -0.114 | -0.014 |
| 150 | 0 | -0.015 | 0.032 | -0.1 | -0.012 |
| 200 | 0 | -0.014 | 0.029 | -0.09 | -0.011 |
| 300 | 0 | -0.011 | 0.024 | -0.074 | -0.009 |
| 400 | 0 | -0.009 | 0.02 | -0.063 | -0.007 |
| 500 | 0 | -0.008 | 0.018 | -0.055 | -0.006 |

Table 6.14: Changes in $1000 \times \ln \mathrm{RPFR}(A^*)$ due to the DBO correction for single carbon-13 substitution. For convenience, the effect is given relative to methane (i.e., the effect on $1000 \times \ln \mathrm{RPFR}(^{13}\mathrm{CH_4})$ is subtracted from each value. "Center," "Terminal," and "Total" refer to the two positions of propane and their appropriate average.

calculated with PIMC are not necessarily more accurate than the harmonic result for several reasons that we discuss at the end of this Chapter. Here we assess the relative importance of the harmonic approximation by comparing the PI-based result relative to the harmonic result using the same description of the molecular potential energy, the F12/ATZ PES.

Anharmonic effects are frequently neglected, since they are not easy to evaluate (especially for larger molecules) and accounting for them typically does not affect the isotopic equilibria dramatically. The typical impact of the anharmonic effects on fractionation (Fig. 6.11), site-specific effects (Fig. 6.12) and clumping (Figs. 6.13-6.14) are shown in the results. The effects shown there vary from rather substantial (e.g., about 20% relative difference for fractionation of deuterium between methane and propane, Fig. 6.11(a)) to completely negligible (e.g., <0.1‰ difference in D + D clumping of methane, Fig. 6.14(a)). In general, the effect on the clumped isotope effect is least significant. For methane it has already been shown that the path-integral-based calculations of clumping agree very well with the harmonic approximation.[86,98] Figs. 6.14 and B.12, B.13, B.14, B.15 show that within the statistical error of the PIMC calculations, this is also true for water, ethane, and propane. It is possible to reduce the statistical uncertainty in clumped isotope effects computed with the path-integral method with additional sampling, but we do not attempt to do it here as preliminary results do not indicate major effect sizes. Therefore, for all but two clumped isotope effects considered here we recommend using the values based on the harmonic frequencies computed with the F12/ATZ method. However, for the two most pronounced (i.e., largest numerical value of $\Delta^{eq}$) clumped isotope effects, which are. D + D in molecular hydrogen and water, we recommend using the path-integral based data, as the anharmonic effects lead to small but verified difference of up to 2‰ for water and up to 4‰ for molecular hydrogen (see Fig. 6.13).

The site-specific isotope effect in propane is affected by about +5‰ for deuterium and -0.3‰ for carbon-13. These effects are larger than the DBO corrections, but less significant than the accurate treatment of molecular potential (Figure 6.4).

Harmonic approximation can yield to significant changes in fractionation equilibria, leading to significant changes in interpretation of observed fractionations. Harmonic calculations overestimate how heavy propane is, which leads to a substantial error for fractionations involving propane (see Fig. 6.11 for the example of methane-propane fractionation, with 20% relative error for deuterium and 5% relative error for carbon-13). More dramatically, the fractionation of deuterium between ethane and propane has a small magnitude and, in this case, harmonic results yield a different sign (i.e., direction) of fractionation as shown in Table 6.15. Harmonic results predict that propane is heavier than ethane with respect to fractionation of deuterium. The full anharmonic (path-integral-based) treatment using the same molecular potential disagrees, making propane lighter than ethane over almost the

| t,°C | Terminal | | Center | | Total | |
|---|---|---|---|---|---|---|
| | PI | Harm | PI | Harm | PI | Harm |
| 0 | -27.1929 | -9.3282 | 79.8337 | 91.371 | 0.6567 | 24.4499 |
| 25 | -25.0998 | -8.7796 | 66.3261 | 77.1366 | -1.4478 | 19.8909 |
| 50 | -23.1613 | -8.2539 | 56.0193 | 65.5601 | -2.7707 | 16.2649 |
| 75 | -21.6203 | -7.7589 | 47.4204 | 56.0636 | -3.9081 | 13.3512 |
| 100 | -20.1859 | -7.2969 | 40.5301 | 48.2133 | -4.6578 | 10.9894 |
| 150 | -17.6816 | -6.4711 | 29.4022 | 36.2036 | -5.7012 | 7.4756 |
| 200 | -15.8421 | -5.7646 | 22.0475 | 27.6798 | -6.2343 | 5.0766 |
| 300 | -12.6429 | -4.6411 | 12.6769 | 16.9318 | -6.2526 | 2.2278 |
| 400 | -10.3284 | -3.8036 | 7.3772 | 10.8864 | -5.8725 | 0.7803 |
| 500 | -8.648 | -3.165 | 4.5265 | 7.2894 | -5.3381 | 0.0263 |

Table 6.15: Comparison of the path-integral-based (PI) and approximate harmonic (Harm) $1000 \times \ln {}^{D}\alpha^{eq}_{\mathrm{propane-ethane}}$ values for the fractionation between propane (as well as its two sites) and ethane. Note that these values differ from those presented in Table 6.6 because the DBO correction has not been applied here.

entire range of temperatures studied here. Note that this is possible in any molecule with large site-specific preference. Harmonic results correctly predicts that ethane is in between the two sites of propane with the terminal site lighter and center site heavier than ethane, but it fails to accurately predict the balance between the two. Also note that this fractionation does not yield a typical temperature dependence: the overall fractionation between ethane and propane now vanishes not only at infinite temperature, but also the second time at approximately 12°C, where the enrichment (depletion) of the center (terminal) sites of propane cancel out precisely with each other. We predict that below this crossover temperature propane becomes heavier than ethane and indeed our DBO-corrected fractionation value $1000 \times \ln {}^{D}\alpha^{eq}_{\mathrm{C_3H_8-C_2H_6}}$ at 0°C is +1.17‰ (see Table 6.6).

We note that the path-integral-based approaches are not the only way to improve on the harmonic approximation. The anharmonic corrections have been applied to isotopic equilibria between alkanes, water and other small molecules.[86,100,130,132] [86] showed that including only the anharmonic corrections to the zero-point energy does not necessarily improve agreement with the path-integral method, particularly they show that for $^{13}$C + D clumped effect in methane the other approximations conspire such that the harmonic result agrees well with the full anharmonic treatment using PIMC. [132] showed that at MP2/aug-cc-pVTZ level of theory, inclusion of many corrections that go beyond harmonic approximation can have a major effect on the isotopic equilibria between molecules. In the subsequent study they established that the effect on clumped isotope effects is substantially weaker.[130] Both of these findings echo our conclusions based on a path-integral-based method and a higher quality (CCSD(T)-F12A/aug-cc-pVTZ) molecular potential. Authors of ref.[100] use a highly accurate CCSD(T)/6–311+G(d,p) potential for calculating harmonic

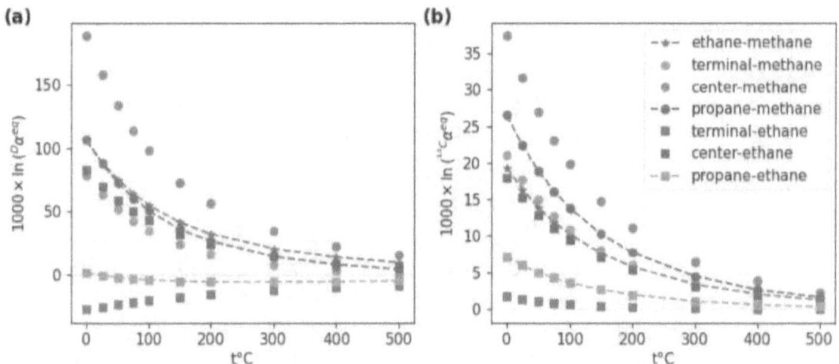

Figure 6.19: **Deuterium (a) and carbon-13 (b) fractionation between alkanes** calculated with PIMC and after applying the DBO corrections. Fractionations of ethane with methane are stars, propane with methane are circles, and propane with ethane are squares. Dashed lines signify total fractionations shown in Table 6.6.

frequencies of butane, but their inclusion of anharmonic effects is based on the lower quality MP2/6–311+G(d,p) potential. The advantage of our approach is that we need not mix different potentials, which can lead to unknown systematic error. On the other hand, this approach is a practical alternative when the path-integral-based methods are not feasible — either because the molecule is too large or because the clumped isotope effects are too weak.

**Trends in heavy isotope fractionation among alkanes**

Figure 6.19 compares the fractionation of deuterium (a) and carbon-13 (b) for the alkanes. Comparing the two, we see many similarities. The center position of propane is the heaviest, while methane is the lightest in both cases. Deuterium fractionation has a significantly larger magnitude and interestingly ethane falls in between the two positions of propane, as discussed in the previous section. This leads to a crossover temperature of about 12°C, below which propane is heavier than ethane and vice versa for the temperatures above 12°C. This is not observed for carbon-13 (compare pink squares on panels (a) and (b) of Fig. 6.19). Deuterium fractionation between ethane and methane is similar to that between propane and methane — but this similarity is the result of averaging between propane's center and terminal sites.

**Confusion over the site-specific effect in propane**

Our definition of $1000 \times \ln \alpha^{eq}$ (Eq. 3.19) to quantify the site-specific isotope effect and $\Delta^{eq}$ (Eq. 3.20) to quantify clumping most closely resembles $\Delta_{K_{eq}^j}$ defined at the end of the methods section in Ref. [87], the only difference being that we use natural logarithm, whereas they subtract one from the ratio of equilibrium constants. Thus,

| t,°C | $1000 \times \ln {}^{D}\alpha^{eq}_{\text{center-terminal}}$ | | | | $1000 \times \ln {}^{13}\alpha^{eq}_{\text{center-terminal}}$ | | | | $\Delta^{eq}_{CH_3-{}^{13}CHD-CH_3}$ | |
|---|---|---|---|---|---|---|---|---|---|---|
| | [87] | [86] | | Here | [87] | [86] | | Here | [87] | Here |
| | | Harm | PIMC | | | Harm | PIMC | | | |
| -23.15 | 122.9 | Harm | PIMC | 130.1 | 18.39 | Harm | PIMC | 19.14 | 6.37 | 6.33 |
| 26.85 | 89.1 | 90.12 | 93.9 | 93.6 | 13.35 | 14.34 | 14.6 | 13.96 | 4.84 | 4.82 |
| 76.85 | 66.4 | | | 71 | 9.78 | | | 10.34 | 3.78 | 3.77 |
| 126.85 | 50.7 | 52.12 | 53.3 | 55.3 | 7.32 | 7.92 | 8.12 | 7.77 | 3.01 | 2.99 |
| 176.85 | 39.5 | | | 43.9 | 5.54 | | | 5.91 | 2.43 | 2.41 |
| 226.85 | 31.4 | 32.93 | 33.6 | 35.5 | 4.25 | 4.6 | 4.63 | 4.54 | 1.97 | 1.96 |

Table 6.16: Comparing $1000 \times \ln \alpha^{eq}$ obtained from Table 3 of Ref. [87], the harmonic and path-integral-based results from Table 3 in Ref. [86] and our best results.

our results presented in Table 6.8 above closely resemble those from Table 3 of Ref. [87], although for proper comparison the values from their Table 3 need to be transformed to $1000 \times \ln \alpha^{eq}$ as follows:

$$1000 \times \ln \alpha^{eq} = 1000 \times \ln \left( \frac{\Delta_{K^j_{eq}}}{1000} + 1 \right). \tag{6.2}$$

We present a head-to-head comparison between our best results (interpolated everywhere except for the lowest temperature point) and Table 3 of Ref. [87] in Table 6.16.

We note that results from [86] shown in Fig. 5 of Ref. [87] are misinterpreted and the apparent disagreement that results from this misinterpretation is not real. Figure 6.20 and Table 6.17 address the two conventions of presenting site-specific isotope effect. The values reported by [86] express enrichment of the center position relative to random distribution of isotopes – the same nomenclature as is typically used for the clumped isotope effect. These values are much smaller than the site preferences. Conversion of $\Delta_i$ values from Table 3 of [86] to the nomenclature we use in this book is done via the following equations:

$$1000 \times \ln {}^{D}\alpha^{eq}_{\text{center-terminal}} = 1000 \times \ln \left( \frac{4}{1 - \Delta_i/3000} - 3 \right) \tag{6.3}$$

$$1000 \times \ln {}^{13}\alpha^{eq}_{\text{center-terminal}} = 1000 \times \ln \left( \frac{3}{1 - \Delta_i/2000} - 2 \right) \tag{6.4}$$

and yields values in excellent agreement with Ref. [87]. The remaining discrepancy in $1000 \times \ln \alpha^{eq}_{\text{center-terminal}}$ between the harmonic result of [86] and [87] can be easily explained by the different molecular potentials used in two studies.

We note that the excellent agreement of the path-integral based results of Ref. [86] with our best results for $1000 \times \ln {}^{D}\alpha^{eq}_{\text{center-terminal}}$ is largely due to lucky combination of opposing factors (the CHARMM force field they used happens to overestimate the result by an amount that is approximately equal to the contribution of the DBO correction we used). The same effects also conspire to reduce apparent difference in the same result for site preference of carbon-13, but this time the cancellation of opposing effects is not as perfect.

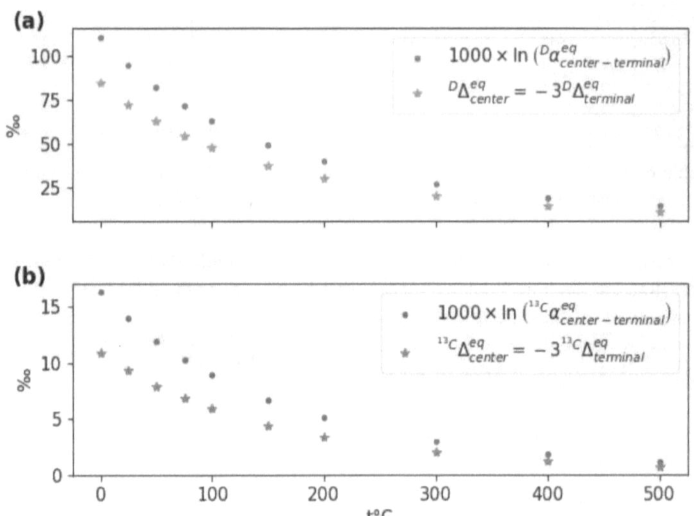


Figure 6.20: **Site-specific effect of propane calculated with the PI method and expressed in two different ways:** as excess heavy atom in the enriched position relative to the depleted (dots) or as excess heavy atom in the enriched position relative to random distribution of heavy isotopes (stars). The side-by-side comparison of the numerical values is in Table 6.17.

| t, °C | $1000 \times \ln{}^D\alpha^{eq}$ | $^D\Delta^{eq}$ | $1000 \times \ln{}^D\alpha^{eq}$ | $^{13}\Delta^{eq}$ |
|---|---|---|---|---|
| 0 | 110.5 | 85.12 | 16.5 | 11.03 |
| 25 | 94.58 | 72.6 | 14.13 | 9.44 |
| 50 | 82.09 | 62.82 | 12.09 | 8.08 |
| 75 | 71.74 | 54.76 | 10.48 | 7 |
| 100 | 63.24 | 48.17 | 9.05 | 6.04 |
| 150 | 49.31 | 37.43 | 6.82 | 4.55 |
| 200 | 39.88 | 30.2 | 5.23 | 3.49 |
| 300 | 26.96 | 20.36 | 3.12 | 2.08 |
| 400 | 19.1 | 14.4 | 1.92 | 1.28 |
| 500 | 14.39 | 10.83 | 1.2 | 0.8 |

Table 6.17: Site-specific isotope effect in propane expressed as both the site preference ($1000 \times \ln \alpha^{eq}$) and the enrichment relative to random distribution of isotopes ($\Delta^{eq}$).

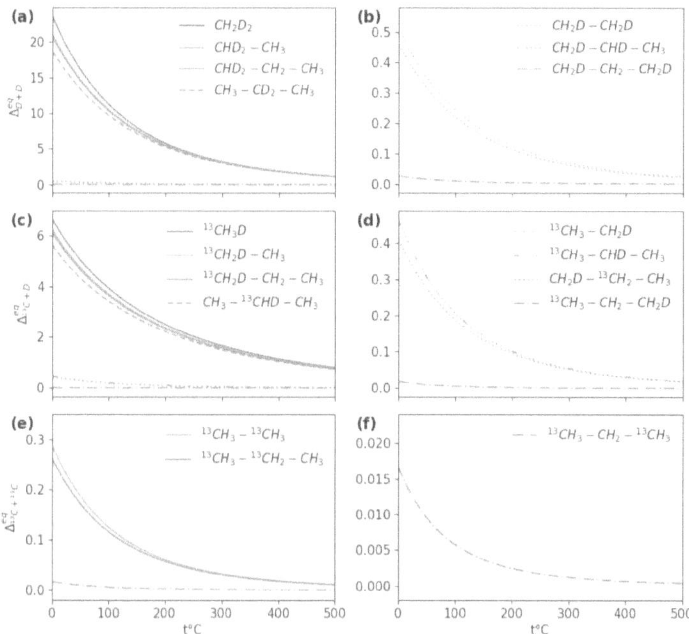

Figure 6.21: **Summary of** D + D **(a-b),** $^{13}$C + D **(c-d) and** $^{13}$C + $^{13}$C **(e-f) clumped isotope effects in alkanes.** The right-hand side panels zoom in to the area around 0 to distinguish weaker clumped effects.

**Trends in clumped isotope effects**

Figure 6.21 summarizes the trends in clumped isotope effects of alkanes. As expected, the D + D clumping is the strongest, followed by $^{13}$C + D clumping with $^{13}$C + $^{13}$C clumping being the weakest. The strength of the clumped isotope effect also strongly depends on the proximity of the heavy isotopes, falling below 1‰ for the deuterium atoms bound to different carbons (panel b) and for $^{13}$C + D clumping when the two heavy atoms are not directly connected to each other (panel d). Further separation of heavy isotopes decreases the effect by another order of magnitude to below 0.1‰. The $^{13}$C + $^{13}$C clumping in propane is similarly affected (panels e-f). There is another less pronounced trend shown on Fig. 6.21. The clumped isotope effects decrease going from methane to primary to secondary carbon for all clumped heavy isotope effects up to propane. Interestingly, propane presents a choice between placing either carbon-13 or deuterium into the primary (i.e., terminal) position. If the two alternatives are compared (Fig. 6.21d), it is clear that carbon-13 has a stronger preference for the primary (terminal) position than deuterium.

| Clumped effect | Methane $\Delta^{eq}$ vs $\Delta$[a] | | Ethane $\Delta^{eq}$ vs $\Delta$ | | Propane $\Delta^{eq}$ vs $\Delta$ | |
|---|---|---|---|---|---|---|
| D + D | 23.7 | 23.53 | 8.74 | 8.63 | 5.33 | 4.82 |
| $^{13}$C + D | 6.64 | 6.55 | 3.35 | 3.26 | 1.98 | 2.1 |
| $^{13}$C + $^{13}$C | N/A | N/A | 0.288 | 0.276 | 0.181 | 0.138 |
| $M + 2$ | 6.99 | 6.9 | 0.751 | 0.724 | 0.391 | 0.342 |

Table 6.18: Comparison of the $\Delta^{eq}$ values used throughout in this study and the corresponding $1000 \times \ln(\Delta/1000 + 1)$ values at 0°C.

## Combined clumped isotope effects

For methane, there is only one doubly substituted isotopologue that determines the total clumped isotope effect for each of the two (D+D and $^{13}$C+D) clumping effects: $CH_2D_2$ determines $\Delta^{eq}_{D+D}$ and $^{13}CH_3D$ determines $\Delta^{eq}_{^{13}C+D}$. However, for ethane and propane the individual clumped effects from Fig. 6.21 can be combined, yielding the total D + D and $^{13}$C + D clumped effects shown in Fig. 6.22. Similarly, for propane the two $^{13}$C + $^{13}$C effects are combined to yield $\Delta^{eq}_{^{13}C+^{13}C}$. Panel (d) of Fig. 6.22 also presents measures of the clumped effects that accommodate low-resolution mass spectroscopy $\Delta^{eq}_{M+2}$, i.e., $\Delta^{eq}_{18}$ for methane and the corresponding values for ethane and propane. All $\Delta^{eq}$ values for combined clumped isotope effects are obtained via simple averaging of relevant $\Delta^{eq}_{\text{isotopologue}}$ values. Fig. 6.22 also compares these averages to the corresponding $\Delta$-values (obtained from Eq. 3.5 and directly relatable to the mass-spectroscopic measurements), obtained from total excess of all relevant species at equilibrium relative to when isotopes are distributed randomly. Recall, that $\Delta^{eq}_i$ values are only equal to $\Delta_i$ values in the limit of infinite dilution, i.e., when heavy isotope abundances are infinitely small. Therefore, even for methane although $\Delta^{eq}_{D+D} = \Delta^{eq}_{CH_2D_2}$ and $\Delta_{D+D} = \Delta_{CH_2D_2}$, but $\Delta^{eq}_{D+D} \neq \Delta_{D+D}$. The situation is analogous for the $^{13}$C + D clumped effect. Both methane and ethane have larger equilibrium values, i.e., $\Delta^{eq} > \Delta$. This is expected as the strength of clumping decreases with increasing concentration of heavy isotopes (see Fig. A.1). The errors are significantly larger for propane, as the naive averaging of $\Delta^{eq}$ (i.e., using weights given by random distribution of isotopologues) does not adequately capture the complexity of equilibrium in propane, that involves the combination of clumped and site-specific effects.

## Rotamers of ethane and propane

Certain experimental techniques (e.g., NMR) allows for resolution of different stable rotational conformers of isotopically substituted ethane and propane, especially at lower temperatures where rotations around C-C single bond are slowed down. For ethane, only the D + D clumped isotope effect on the neighboring carbons results in two different rotamers, shown in Fig. 6.23(a-b). The clumped isotope effect for

---

[a]For proper comparison to $\Delta^{eq}$, the numerical values of $\Delta$ from Eq. 3.5 are given as $1000 \times \ln(\Delta/1000 + 1)$.

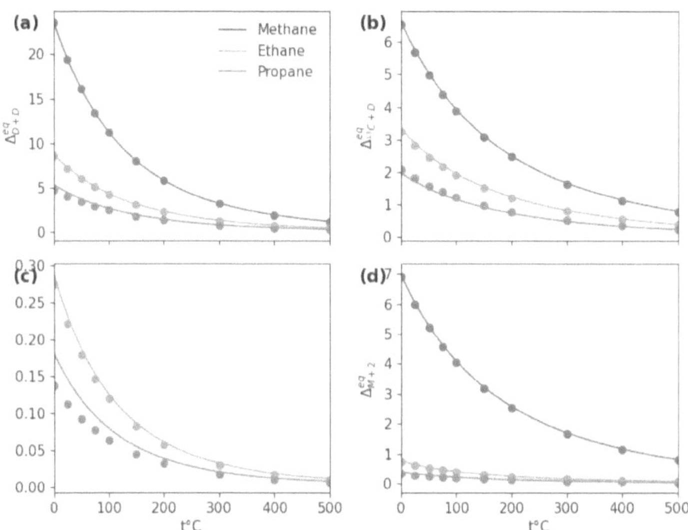

Figure 6.22: **Combined** $D + D$ **(a),** $^{13}C + D$ **(b) and** $^{13}C + ^{13}C$ **(c) clumped isotope effects for alkanes.** Panel (d) presents a combination of all the clumped effects. The $\Delta^{eq}$ values (solid lines) are compared to the $1000 \times \ln(\Delta/1000 + 1)$ values (circles) as defined in Eq. 3.5. Table 6.18 includes a head-to-head comparison of these values at 0°C.

| Rotamers of | Larger clumping rotamer | | Smaller clumping rotamer | |
|---|---|---|---|---|
| $CH_2D - CH_2D$ | (a) D *trans* to D | 1.33‰ | (b) D *gauche* to D | 0.10‰ |
| $CH_2D - CHD - CH_3$ | (c) D *trans* to D | 1.23‰ | (d) D *gauche* to D | 0.09‰ |
| $CHD_2 - CH_2 - CH_3$ | (e) H *trans* to Me[b] | 21.04‰ | (f) D *trans* to Me | 20.68‰ |
| $^{13}CH_2D - CH_2 - CH_3$ | (g) D *trans* to Me | 6.13‰ | (h) H *trans* to Me | 6.06‰ |
| $CH_2D - ^{13}CH_2 - CH_3$ | (h) H *trans* to Me | 0.45‰ | (g) D *trans* to Me | 0.33‰ |
| $CH_2D - CH_2 - ^{13}CH_3$ | (g) D *trans* to Me | 0.043‰ | (h) H *trans* to Me | 0.0044‰ |

Table 6.19: Clumped isotope effects ($\Delta^{eq}$) in rotamers of ethane and propane at 0°C with F12/ATZ harmonic frequencies. The rotamer labels (a)-(h) refer to Fig. 6.23

rotamer (a) with deuterium atoms at 180°angle relative to each other is about 10 times larger than for the rotamer shown in (b). For the temperature ranges considered here these rotamer freely interconvert and so elsewhere we report only the average (properly weighted as 1 to 2) for the two rotamers. The difference of about a factor of 10 is also observed for an analogous situation for both $D + D$ (Fig. 6.23c-d) and $^{13}C + D$ (Fig. 6.23g-h) clumped effects in propane. All other clumped isotope effects of propane are much closer to each other in numerical value for each of two rotamers (see Table 6.19 for comparison at 0°C).

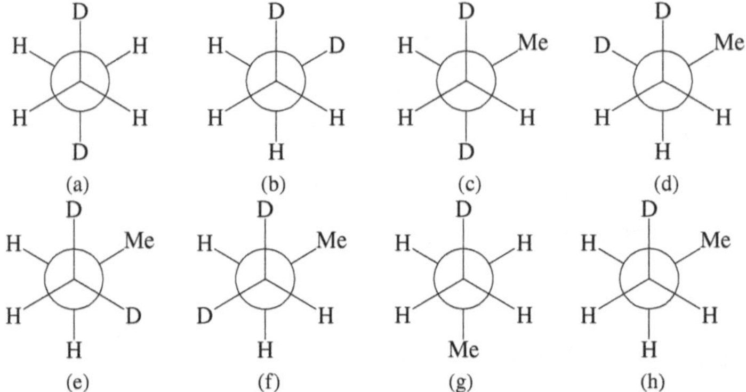

(a)        (b)        (c)        (d)

(e)        (f)        (g)        (h)

Figure 6.23: **Rotamers of doubly substituted ethane (a-b) and propane (c-h).** Carbon-13 is not labelled, so (g) and (h) describe all relevant $^{13}C + D$ clumped effects. In each case the front carbon of the Newman projection corresponds to the first carbon in Table 6.19.

The terminal site of propane also has two non-equivalent positions for the substitution of H by D as shown in the Newman projections (g) with D *trans* to the methyl group and (h) with D *gauche* to the methyl group. We do not observe a significant difference in RPFR for the two either from the path-integral or from the harmonic calculations.

**Considerations for the Path-integral calculations**

Properly done PI-based calculations yield accurate RPFR's for a particular potential. This means that when low-quality affordable potentials are used, we cannot expect accurate predictions for RPFR's. In particular, the prior studies of position-specific isotope effect in propane[86,102] should not be treated as reference values for the experimental predictions, but only as the estimates of the effect of anharmonicities for the potentials the path-integral-based approach was applied to. It is possible that the more accurate potentials have similar anharmonic effects, but we have not investigated whether this is the case for the molecules investigated here, and we do not know of a study that did.

We wish to share a note of caution with the reader: the PI calculations are difficult to converge for the "weak" clumped isotope effects, i.e., those whose enrichment is very slightly above randomness. For the molecules studied here the following clumped isotope effects were found to be weak: $^{13}C + ^{13}C$ clumping in ethane and propane as well as D + D clumping on different carbon atoms in ethane and propane as well as $^{13}C + D$ clumping for deuterium atom that is not connected to carbon-13. We initially set out to calculate all of these clumped isotope effects using

---

$^{b}$Me stands for methyl group, i.e., $CH_3$.

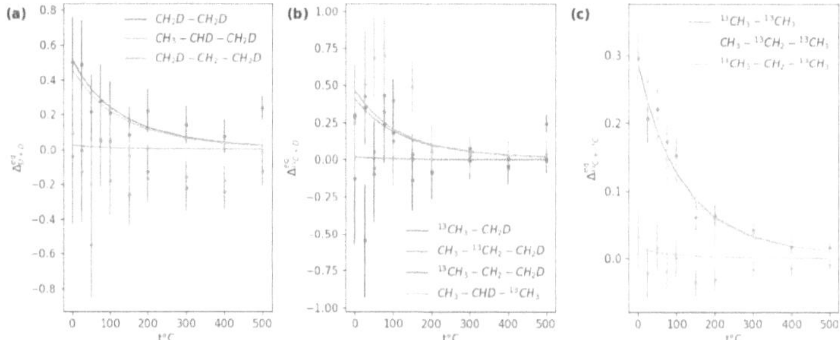

Figure 6.24: **Comparison of the clumped isotope effects in ethane and propane calculated with PIMC and the corresponding harmonic results** for (a) D + D, (b) $^{13}C$ + D and (c) $^{13}C$ + $^{13}C$ clumped effects. The error bars overwhelm the result, requiring at least 100 times more computing time for any meaningful interpretation. These (unconverged) results give no indication that the anharmonic effects are important in these cases.

our standard path-integral-based methodology, but quickly realized that calculating them to a reasonable level of precision would require a huge amount of computing time. Figure 6.24 shows the results with statistical uncertainties after collecting a little over 10 million samples (which is comparable to what we have done for all other clumped effects, see Table 6.2). The issue is that the standard error of $\Delta^{eq}$ does not decrease as the clumped effect is weakened and thus the statistical uncertainty (noise) overwhelms the clumped isotope effect (the signal) making the path-integral approach we used in this study impractical. In these cases, we recommend using the harmonic results To summarize, the PI approach is limited in two ways: (i) The high-level calculations to accurately fit the PES become not feasible for larger and less symmetric molecules. Even derivatives of propane (e.g., propanol) would be challenging. (ii) The sampling to achieve high precision of the target RPFR is costly, especially for small clumped isotope effects.

## 6.5 Summary

Based on the extensive calculations performed in this study, we briefly state our key observations for heavy isotope fractionation, clumping and position-specific effect in gaseous dihydrogen, water, and small alkanes at 0 to 500 degrees C. We believe that these ideas extend beyond the strict confines of the examples considered here and we hope that they will serve a useful guide for future studies of stable isotope geochemistry.

1. Clumped heavy isotope effects enjoy significant error cancellation and can be computed better than experimental accuracy using a simple Bigeleisen-Urey harmonic approach and a cheap DFT potential like B3LYP/6-311G(d,p).

2. On the contrary, fractionation of heavy isotopes between alkanes and position-specific isotope effect in propane require the use of highly accurate potential and accounting for the anharmonic effects.

3. The DBO correction is negligible for equilibria involving carbon-13, but can be important for fractionations and position-specific effects involving deuterium.

4. Clumped heavy isotope effect is the strongest for two atoms that are directly bonded to each other and decays significantly with each additional bond.

5. In alkanes, both D+D and $^{13}$C+D clumped heavy isotope effects of a particular carbon decreases in the following order: $CH_4$ > $CH_3$ > $CH_2$. Moreover, the total clumping decreases sharply due to the increasing importance of the weakly-clumped isotopologue with distant heavy isotopes.

6. Path-integral-based methods are a method of choice for accurate calculations of fractionation between small gaseous molecules. We do not see evidence that path-integral-based approaches are necessary to accurately describe the clumped isotope effects in alkanes.

# DETAILS OF METHODOLOGY

## A.1   ORCA calculations

Our typical ORCA input file is as follows:

```
! CCSD(T)  aug-cc-pVTZ  defgrid3  Opt  NumFreq  Mass2016
VeryTightSCF  NoFrozenCore

* xyzfile  0  1  init.xyz
%freq  Increment  0.001  end
%geom  Convergence  tight  end
%scf  Convergence  VeryTight  end
```

First, the molecular geometry is optimized with tight convergence criteria using CCSD(T)/ATZ method and without using the frozen core approximation starting from the initial guess provided in the "init.xyz" file. The initial molecular geometry is supplied in the "init.xyz" file and comes from optimized geometry given by the cheaper method. As instructed by the keyword "Mass2016," the program uses atomic masses of most abundant isotopes (average atomic masses are used by default). The finite difference method (keyword "NumFreq") is used with the step size or "Increment" of 0.001 Bohr. Harmonic frequencies calculation via finite difference is the most time-consuming part of the entire calculation, accounting for >95% of the total runtime. The frequencies are printed in the ".hess" output file. We note that there is no need to recalculate the harmonic frequencies for each isotopologue from scratch; instead the hessian output file "some_output.hess" can be redirected to the "orca_vib" helper program, that is part of the ORCA package. This will instantly print out the harmonic frequencies of the isotopologue specified in the "$atoms" section of the hessian file.

Harmonic frequency calculations proceed through (i) molecular geometry is optimization (i.e., the minimum energy point search) followed by (ii) forces evaluations at each of 6N small displacements around this equilibrium point to get the Hessian matrix. Following this, (iii) the 3N-by-3N Hessian matrix is diagonalized and weighted by the atomic masses, which gives the harmonic frequencies as the 3N-6 (3N-5 for linear molecules) eigenvalues. The molecular geometry is optimized using a quasi-Newton update procedure with the BFGS update (Eckert et al., 1998). Since the algorithm finds a local minimum, it is crucial for the optimization to start with the geometry that is close to the global minimum in energy. The initial guess is obtained from experiments or a previous calculation (frequently using a less accurate, but also less computationally expensive method) and is guided to the minimum energy geometry using the gradient of energy. We have used the experimentally available structures as a guess for the cheapest RHF/DZ calculations of each molecule and the final optimized geometry served as initial guess for the next (more expensive) method.

It is worth noting that hydrogen atoms, attached to one heavy atom (oxygen or carbon) are equivalent by construction for the PIP surfaces described in section 3.3.2. Thus, substituting either of the two hydrogens in water or either of the four hydrogens in methane gives the same exact harmonic frequencies for the PIP surfaces. This is not the case for the calculations that are typically performed with a standard software package (Gaussian, ORCA, or MolPro), unless molecular symmetry is enforced. Small asymmetries in optimized structure will result in slight differences in harmonic frequencies when different formally equivalent atoms are substituted. To minimize the error introduced by these small asymmetries, we average over the RPFR's due to equivalent substitutions for all of our ORCA calculations.

### A.2  Heavy isotope fractionations involving propane

Using Eq. (3.21), we can directly calculate the bulk fractionation factors $\alpha^{eq}$ for molecules in which all sites that allow for isotopic substitution are equivalent. This is the case for all molecules considered here except for propane, which has two non-equivalent positions for both carbon and hydrogen — the terminal (methyl group) and the center (methylene group). In this case we must write down two heavy isotope exchange reactions, that both contribute to the bulk fractionation of either deuterium or carbon-13 between propane and the other molecule $M$. Consider, for example bulk fractionation of propane with dihydrogen:

$$HD + C_3H_8 \rightleftharpoons H_2 + CH_3 - CH_2 - CH_2D, {}^D\alpha^{eq}_{\text{terminal}-H_2} = \frac{RPFR(CH_3 - CH_2 - CH_2D)}{RPFR(HD)}$$

$$HD + C_3H_8 \rightleftharpoons H_2 + CH_3 - CHD - CH_3, {}^D\alpha^{eq}_{\text{center}-H_2} = \frac{RPFR(CH_3 - CHD - CH_3)}{RPFR(HD)}.$$

Both equilibria contribute to the bulk heavy isotope fractionation with the individual contribution weighted by the number of atoms of this type:

$$^D\alpha^{eq}_{\text{propane}-M} = \frac{3}{4}{}^D\alpha^{eq}_{\text{terminal}-M} + \frac{1}{4}{}^D\alpha^{eq}_{\text{center}-M} \tag{A.1}$$

$$^{13}\alpha^{eq}_{\text{propane}-M} = \frac{2}{3}{}^{13}\alpha^{eq}_{\text{terminal}-M} + \frac{1}{3}{}^{13}\alpha^{eq}_{\text{center}-M} \tag{A.2}$$

While the weights of the individual $\alpha^{eq}$ given in Eqs. A.1-A.2 are approximate, we confirm that this approximation is negligibly different from the exact answer when abundances of heavy isotopes is small (i.e., at close to natural abundances), see Appendix A.3.

### A.3 Approximating isotope fractionations by considering singly substituted isotopologues only

It is well established that singly substituted species play a dominant role in exchange of heavy isotopes between phases. Intuitively this is because (i) the total concentration of all multiply substituted isotopologues (i.e., all isotopologues other than the one with all light isotopes or a single heavy isotope) is small for most small molecules at natural abundance of isotopes (excluding molecules containing chlorine, bromine and other atoms with large abundance of heavy isotopes) and (ii) the deviations from random distribution of multiply substituted isotopologues are typically not very large.

We therefore obtain $\alpha_{M-N}^{eq}$ values that describe fractionations between molecules or sites $M$ and $N$ from equilibrium constants of the exchange reactions between singly substituted isotopologues only, as discussed in Section 3.3. To quantitatively assess whether $\alpha_{M-N}^{eq}$ values (defined in Eq. 3.19) approximate corresponding exact $\alpha_{M-N}$ values (defined in Eq. 3.3) at close to natural abundances, we equilibrate each of the molecules studied here with methane, whose deuterium abundance is set to VSMOW, i.e., 155.76 ppm[211] and whose carbon-13 abundance is set to VPDB i.e., 11.18 ppt.[82] For methane-water equilibrium we additionally set the abundance of oxygen-17 and oxygen-18 as 379.9 ppm and 2005.20 ppm, respectively.[212] The equilibration is done based on the equilibrium constants calculated with our F12/ATZ reference molecular potentials (Section 6.3). The equilibrium constants for isotope exchange reactions between singly substituted isotopologues are given by path-integral-based calculations, while all other equilibrium constants (involving double, triple etc. substitutions) are calculated using the harmonic approach.

Table A.1 below presents comparison between $1000 \times \ln \alpha_{\mathrm{CH_4}-M}$ values (Eq. 3.3) and the corresponding $1000 \times \ln \alpha_{\mathrm{CH_4}-M}^{eq}$ values (Eq 3.19). The total $\alpha^{eq}$ for propane (last row) is obtained by averaging as described in A.2. The total $\alpha$ is obtained as follows for fractionation of deuterium:

$$^{D}\alpha_{\mathrm{CH_4}-M} = \frac{[\mathrm{D}]_{\mathrm{CH_4}}/[\mathrm{H}]_{\mathrm{CH_4}}}{[\mathrm{D}]_M/[\mathrm{H}]_M} \tag{A.3}$$

with $[A]_M$ is the concentration of isotope $A$ in species $M$. The analogous expression is used for carbon-13.

From Table A.1 we conclude that for the isotope fractionations considered here we may neglect any error resulting from assuming $\alpha_{M-N}^{eq} = \alpha_{M-N}$ as they are less than 0.1‰ for the fractionations with methane at 0°C and even smaller at higher temperature.

### A.4 Clumped heavy isotope effect equilibrium reactions

Equation (3.12) directly defines the heavy isotope clumped effect $\Delta_{\mathrm{D+D}}$ for dihydrogen, water and methane, $\Delta_{\mathrm{^{13}C+D}}$ for methane and $\Delta_{\mathrm{^{17}O+D}}$ as well as $\Delta_{\mathrm{^{18}O+D}}$ for water.

---

[a]Values are given as $1000 \times \ln \alpha_{\mathrm{CH_4}-M}$ in all cases.

| $M$ | Fractionation of deuterium | | Fractionation of carbon-13 | |
|---|---|---|---|---|
| | $^D\alpha^a$ | $^D\alpha^{eq}$ | $^{13}\alpha$ | $^{13}\alpha^{eq}$ |
| dihydrogen | 1299.5 | 1299.43 | | |
| water | -142.92 | -142.98 | | |
| ethane | -101.6 | -101.59 | -19.349 | -19.346 |
| term. propane | -74.388 | -74.392 | -20.992 | -20.99 |
| cent. propane | -181.42 | -181.42 | -37.499 | -37.494 |
| total propane | -102.24 | -102.23 | -26.524 | -26.522 |

Table A.1: Comparison between $\alpha$ and $\alpha^{eq}$ (both expressed as $1000 \times \ln \alpha$) for deuterium and carbon-13 fractionation at 0°C. Note that these are prior to the DBO correction, thus numbers shown here differ slightly from our best results presented in Tables 6.6 and 6.7, respectively.

The corresponding equilibria are listed below:

$$\Delta^{eq}_{D+D}(H_2) = \Delta^{eq}_{D_2} : \quad 2\,HD \rightleftharpoons H_2 + D_2 \tag{A.4}$$

$$\Delta^{eq}_{D+D}(H_2O) = \Delta^{eq}_{D_2O} : \quad 2\,HDO \rightleftharpoons H_2O + D_2O \tag{A.5}$$

$$\Delta^{eq}_{^{17}O+D}(H_2O) = \Delta^{eq}_{HD^{17}O} : \quad HDO + H_2{}^{17}O \rightleftharpoons H_2O + HD^{17}O \tag{A.6}$$

$$\Delta^{eq}_{^{18}O+D}(H_2O) = \Delta^{eq}_{HD^{18}O} : \quad HDO + H_2{}^{18}O \rightleftharpoons H_2O + HD^{18}O \tag{A.7}$$

$$\Delta^{eq}_{D+D}(CH_4) = \Delta^{eq}_{CH_2D_2} : \quad 2\,CH_3D \rightleftharpoons CH_4 + CH_2D_2 \tag{A.8}$$

$$\Delta^{eq}_{^{13}+D}(CH_4) = \Delta^{eq}_{^{13}CH_3D} : \quad CH_3D + {}^{13}CH_4 \rightleftharpoons CH_4 + {}^{13}CH_3D. \tag{A.9}$$

Ethane is described by Equations 3.14 and 3.15. We spell out the five individual equilibrium reactions below:

$$\Delta^{eq}_{^{13}C+^{13}C}(C_2H_6) = \Delta^{eq}_{^{13}CH_3 - {}^{13}CH_3} :$$
$$2\,CH_3 - {}^{13}CH_3 \rightleftharpoons C_2H_6 + {}^{13}CH_3 - {}^{13}CH_3 \tag{A.10}$$

$$\Delta^{eq}_{CHD_2 - CH_3} : \quad 2\,CH_2D - CH_3 \rightleftharpoons C_2H_6 + CHD_2 - CH_3 \tag{A.11}$$

$$\Delta^{eq}_{CH_2D - CH_2D} : \quad 2\,CH_2D - CH_3 \rightleftharpoons C_2H_6 + CH_2D - CH_2D \tag{A.12}$$

$$\Delta^{eq}_{^{13}CH_2D - CH_3} : \quad CH_2D - CH_3 + {}^{13}CH_3 - CH_3 \rightleftharpoons C_2H_6 + {}^{13}CH_2D - CH_3 \tag{A.13}$$

$$\Delta^{eq}_{CH_2D - {}^{13}CH_3} : \quad CH_2D - CH_3 {}^+ {}^{13}CH_3 - CH_3 \rightleftharpoons C_2H_6 + CH_2D - {}^{13}CH_3. \tag{A.14}$$

Finally, the clumped isotope effect in propane is described by the following 11 reactions:

$$\Delta^{eq}_{^{13}CH_3 - CH_2 - {}^{13}CH_3}: \quad 2\,{}^{13}CH_3 - CH_2 - CH_3 \rightleftharpoons$$
$$C_3H_8 + {}^{13}CH_3 - CH_2 - {}^{13}CH_3 \tag{A.15}$$

$$\Delta^{eq}_{^{13}CH_3 - {}^{13}CH_2 - CH_3}: \quad {}^{13}CH_3 - CH_2 - CH_3 + CH_3 - {}^{13}CH_2 - CH_3 \rightleftharpoons$$
$$C_3H_8 + {}^{13}CH_3 - {}^{13}CH_2 - CH_3 \tag{A.16}$$

$$\Delta^{eq}_{^{13}CH_2D - CH_2 - CH_3}: \quad {}^{13}CH_3 - CH_2 - CH_3 + CH_2D - CH_2 - CH_3 \rightleftharpoons$$
$$C_3H_8 + {}^{13}CH_2D - CH_2 - CH_3 \tag{A.17}$$

$$\Delta^{eq}_{CH_2D - {}^{13}CH_2 - CH_3}: \quad CH_3 - {}^{13}CH_2 - CH_3 + CH_2D - CH_2 - CH_3 \rightleftharpoons$$
$$C_3H_8 + CH_2D - {}^{13}CH_2 - CH_3 \tag{A.18}$$

$$\Delta^{eq}_{CH_2D - CH_2 - {}^{13}CH_3}: \quad {}^{13}CH_3 - CH_2 - CH_3 + CH_2D - CH_2 - CH_3 \rightleftharpoons$$
$$C_3H_8 + CH_2D - CH_2 - {}^{13}CH_3 \tag{A.19}$$

$$\Delta^{eq}_{CH_3 - {}^{13}CHD - CH_3}: \quad CH_3 - {}^{13}CH_2 - CH_3 + CH_3 - CHD - CH_3 \rightleftharpoons$$
$$C_3H_8 + CH_3 - {}^{13}CHD - CH_3 \tag{A.20}$$

$$\Delta^{eq}_{^{13}CH_3 - CHD - CH_3}: \quad {}^{13}CH_3 - CH_2 - CH_3 + CH_3 - CHD - CH_3 \rightleftharpoons$$
$$C_3H_8 + {}^{13}CH_3 - CHD - CH_3 \tag{A.21}$$

$$\Delta^{eq}_{CHD_2 - CH_2 - CH_3}: \quad 2\,CH_2D - CH_2 - CH_3 \rightleftharpoons$$
$$C_3H_8 + CHD_2 - CH_2 - CH_3 \tag{A.22}$$

$$\Delta^{eq}_{CH_3 - CD_2 - CH_3}: \quad 2\,CH_3 - CHD - CH_3 \rightleftharpoons$$
$$C_3H_8 + CH_3 - CD_2 - CH_3 \tag{A.23}$$

$$\Delta^{eq}_{CH_2D - CH_2 - CH_2D}: \quad 2\,CH_2D - CH_2 - CH_3 \rightleftharpoons$$
$$C_3H_8 + CH_2D - CH_2 - CH_2D \tag{A.24}$$

$$\Delta^{eq}_{CH_2D - CHD - CH_3}: \quad CH_2D - CH_2 - CH_3 + CH_3 - CHD - CH_3 \rightleftharpoons$$
$$C_3H_8 + CH_2D - CHD - CH_3. \tag{A.25}$$

## A.5  Relationship between $\Delta$ and $\Delta^{eq}$

Throughout the study we report the $\Delta^{eq}$ values defined in Eq. (3.20). These are independent of heavy isotope abundances and express only the thermodynamic preference for two heavy isotopes to occupy the same site of the molecule at a given temperature. In contrast, experimentally measured $\Delta$ values (defined in Eq. 3.5) depend on isotopic abundance and are relatable to the corresponding $\Delta^{eq}$ values only in the limit of infinite dilution:

$$\lim_{[D]\to 0,\, [^{13}C]\to 0} \left\{ 1000 \times \ln\left(\frac{\Delta}{1000} + 1\right) \right\}. \tag{A.26}$$

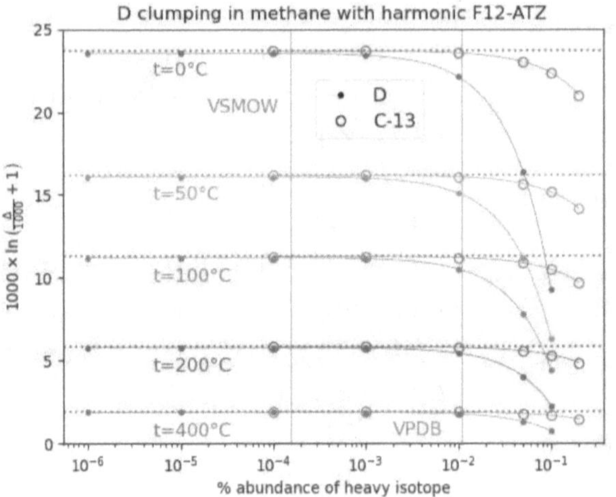


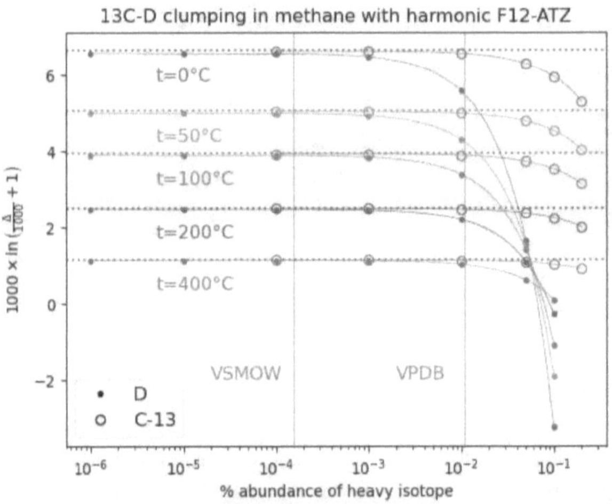


Figure A.1: **Effect of the abundance of heavy isotopes on (a)** D + D **and (b)** $^{13}$C + D **clumped isotope effects in methane.** Dotted horizontal lines are concentration-independent $\Delta^{eq}$ values (i.e., infinite dilution limit, see Eq. A.26). Vertical lines label the VPDB and VSMOW standards for carbon-13 and deuterium, respectively. When abundance of deuterium is varied (dots), carbon-13 is set to VPDB; when abundance of carbon-13 is varies (open circles), deuterium is set to VSMOW.

| [D] | $1000 \times \ln\left(\Delta_{CH_2D_2}/1000 + 1\right)$ | | | | | $1000 \times \ln\left(\Delta_{^{13}CH_3D}/1000 + 1\right)$ | | | | |
|---|---|---|---|---|---|---|---|---|---|---|
| | 0°C | 50°C | 100°C | 200°C | 400°C | 0°C | 50°C | 100°C | 200°C | 400°C |
| $\Delta^{eq}$ | 23.7 | 16.177 | 11.284 | 5.829 | 1.909 | 6.637 | 5.048 | 3.931 | 2.503 | 1.142 |
| $10^{-6}$ | 23.552 | 16.064 | 11.197 | 5.773 | 1.884 | 6.563 | 4.992 | 3.888 | 2.476 | 1.129 |
| $10^{-5}$ | 23.551 | 16.064 | 11.197 | 5.773 | 1.883 | 6.562 | 4.992 | 3.887 | 2.475 | 1.129 |
| $10^{-4}$ | 23.538 | 16.055 | 11.191 | 5.77 | 1.882 | 6.553 | 4.985 | 3.883 | 2.473 | 1.128 |
| $10^{-3}$ | 23.409 | 15.967 | 11.13 | 5.738 | 1.872 | 6.465 | 4.923 | 3.838 | 2.448 | 1.119 |
| $10^{-2}$ | 22.134 | 15.1 | 10.526 | 5.428 | 1.771 | 5.592 | 4.311 | 3.397 | 2.204 | 1.028 |
| 0.05 | 16.758 | 11.441 | 7.979 | 4.115 | 1.342 | 1.934 | 1.743 | 1.548 | 1.182 | 0.647 |
| 0.1 | 10.68 | 7.296 | 5.09 | 2.625 | 0.855 | -2.155 | -1.128 | -0.518 | 0.04 | 0.222 |

Table A.2: Dependence of clumped isotope in methane on abundance of D, while keeping $^{13}$C at VPDB (F12/ATZ potential used).

| [$^{13}$C] | $1000 \times \ln\left(\Delta_{CH_2D_2}/1000 + 1\right)$ | | | | | $1000 \times \ln\left(\Delta_{^{13}CH_3D}/1000 + 1\right)$ | | | | |
|---|---|---|---|---|---|---|---|---|---|---|
| | 0°C | 50°C | 100°C | 200°C | 400°C | 0°C | 50°C | 100°C | 200°C | 400°C |
| $\Delta^{eq}$ | 23.7 | 16.177 | 11.284 | 5.829 | 1.909 | 6.637 | 5.048 | 3.931 | 2.503 | 1.142 |
| $10^{-4}$ | 23.676 | 16.16 | 11.273 | 5.823 | 1.907 | 6.621 | 5.037 | 3.923 | 2.499 | 1.14 |
| $10^{-3}$ | 23.664 | 16.151 | 11.266 | 5.818 | 1.905 | 6.615 | 5.032 | 3.919 | 2.497 | 1.139 |
| $10^{-2}$ | 23.544 | 16.06 | 11.195 | 5.773 | 1.884 | 6.555 | 4.987 | 3.884 | 2.474 | 1.129 |
| 0.05 | 23.012 | 15.655 | 10.88 | 5.573 | 1.793 | 6.289 | 4.785 | 3.727 | 2.374 | 1.083 |
| 0.1 | 22.347 | 15.15 | 10.487 | 5.322 | 1.679 | 5.956 | 4.532 | 3.53 | 2.249 | 1.026 |
| 0.2 | 21.017 | 14.139 | 9.7 | 4.821 | 1.45 | 5.291 | 4.026 | 3.136 | 1.998 | 0.912 |

Table A.3: Dependence of clumped isotope in methane on abundance of $^{13}$C, while keeping D at VSMOW (F12/ATZ potential used).


Fig. A.1 addresses the effect of heavy isotope abundance on the strength of experimentally measured excess of doubly substituted species (i.e., $\Delta$ values) in methane. The values plotted are mirrored in the Tables A.2 and A.3. At low (near-natural) abundances the two are close to each other, but increasing the abundance of heavy isotopes decreases the strength of measured clumping. Notably, at surprisingly low abundance of deuterium of 10% the apparent $^{13}$C + D clumped isotope effect reverses direction, i.e., the doubly substituted $^{13}CH_3D$ species is less abandoned than at random distribution of isotopes as evidenced by the negative values in the last rows of Table A.2. This is because the D + D clumped isotope effect in methane is significantly stronger than the $^{13}$C + D clumping, so it leads to the depletion of singly deuterated methane that is large enough to force "reverse clumping" of $^{13}$C + D.

This is made more concrete by Table A.4. At low abundance of deuterium, $\Delta_{CH_3D}$ is tiny, so the two clumped isotope effects are essentially independent with each of them close to the infinite dilution limit. However, at high deuterium abundance the $D + D$ clumped isotope effect leads to significant depletion of $CH_3D$ with the $\Delta_{CH_3D}$ comparable in size to $\Delta^{eq}_{^{13}CH_3D}$. The isotopologues with carbon-13 and more than one deuterium substitution also deplete the equilibrium amount of $^{13}CH_4$. As a result, at equilibrium the denominator of the right-hand side of the Eq. A.27 is sufficiently small that the numerator needs to be smaller than unity to obtain the

| [D] | $CH_3D$ | $CH_2D_2$ | $CHD_3$ | $CD_4$ | $^{13}CH_4$ | $^{13}CH_3D$ | $^{13}CH_2D_2$ | $^{13}CHD_3$ | $^{13}CD_4$ |
|---|---|---|---|---|---|---|---|---|---|
| $10^{-5}$ | -0.074 | 23.83 | 48.31 | 73.37 | -0.00027 | 6.58 | 30.65 | 55.29 | 80.52 |
| 0.1 | -6.49 | 10.74 | 28.26 | 46.09 | -2.28 | -2.15 | 15.15 | 32.75 | 50.66 |

Table A.4: $\Delta$ values of all isotopologues of methane at low vs high abundance of deuterium, both at 0°C.

left-hand side of the equation, which is only very slightly larger than one.

$$1 \lesssim \frac{K_{^{13}CH_3D}}{K_{^{13}CH_3D}^{\mathrm{random}}} = \frac{\Delta_{^{13}CH_3D} + 1}{\left(\Delta_{CH_3D}\right) \times \left(\Delta_{^{13}CH_4}\right)} \tag{A.27}$$

Propane is distinct compared to other molecules here for clumped isotopes because it has two non-equivalent sites for placing carbon-13 or deuterium. The presence of these non-equivalent sites leads to significant deviations from a random distribution for the equilibrium concentrations of singly-substituted isotopologues of propane at each site. This is distinct from other molecules where, at natural abundances, the concentration of the single substituted isotopologues is effectively the same as that for the random distribution. As a result, the $\Delta$ values for clumping on the individual sites of propane represent a combination of the clumped and position-specific heavy isotope effects. To be more specific, the $\Delta$ value for clumping in the methylene group of propane is significantly larger than expected based on the clumped heavy isotope effect considerations alone, while the same value for the methyl group is negative. Both of these are due to the site-specific heavy isotope effect in propane, which makes the methylene group significantly heavier than the methyl group. In contrast, $\Delta^{eq}$ describe the heavy isotope clumped effect alone. These values are always positive and have similar magnitudes to analogous reactions for methane and ethane – provided one takes care to consider the equilibrium reactions that do not involve shifts in heavy isotope positions.

Table A.5 relates the $\Delta^{eq}$ values for up to doubly substituted isotopologues of propane to experimentally measurable $\Delta$ values (calculated with carnon-13 at VPDB and deuterium at VSMOW abundances).[82,211] For propane, the $\Delta^{eq}$ values do not approximate $1000 \times \ln(\Delta/1000 + 1)$ values even in the limit of infinite dilution, other than for singly substituted species. This is because the $\Delta^{eq}$ value of each clumped isotopologue decouples clumping from the site preference and reports the thermodynamic preference of clumping. In contrast, the experimentally measured $\Delta$ values combine the (strong) site preference and the (much weaker) clumped effect. Thus, the $\Delta$ values and corresponding $1000 \times \ln(\Delta/1000 + 1)$ values are negative for all doubly substituted $D + D$ and $^{13}C + D$ isotopologues where no deuterium atom is in the center position. As suggested by [91], addition of the corresponding pair of $\Delta^{eq}$ of singly substituted isotopologues recovers the $1000 \times \ln(\Delta/1000 + 1)$ as a crude approximation (see second to last column of the table).

| Species | Heavy atoms | $[A]$ | $[A]_{\text{random}}$ | $1000 \times \ln\left(\frac{\Delta}{1000} + 1\right)$ | $\Delta_2^{eq} + 2\Delta_1^{eq}$ | $\Delta^{eq}$ |
|---|---|---|---|---|---|---|
| $CH_3 - CH_2 - CH_3$ | None | 0.965992 | 0.965992 | 0 | | 0 |
| $CH_2D - CH_2 - CH_3$ | D | 0.000879 | 0.000903 | -26.22 | | -28.37 |
| $CH_3 - CHD - CH_3$ | D | 0.000324 | 0.000301 | 74.48 | | 81.69 |
| $CHD_2 - CH_2 - CH_3$ | D&D | 1.36E-07 | 1.41E-07 | -31.64 | -35.94 | 20.8 |
| $CH_3 - CD_2 - CH_3$ | D&D | 2.77E-08 | 2.34E-08 | 167.7 | 182.1 | 18.7 |
| $CH_2D - CHD - CH_3$ | D&D | 2.95E-07 | 2.81E-07 | 48.72 | 53.77 | 0.4594 |
| $CH_2D - CH_2 - CH_2D$ | D&D | 2.00E-07 | 2.11E-07 | -52.41 | -56.72 | 0.0265 |
| $^{13}CH_3 - CH_2 - CH_3$ | $^{13}C$ | 0.021477 | 0.0216 | -5.672 | N/A | -5.516 |
| $CH_3 - ^{13}CH_2 - CH_3$ | $^{13}C$ | 0.010922 | 0.0108 | 11.23 | N/A | 10.97 |
| $^{13}CH_3 - ^{13}CH_2 - CH_3$ | $^{13}C$ & $^{13}C$ | 0.000243 | 0.000241 | 5.821 | 5.719 | 0.263 |
| $^{13}CH_3 - CH_2 - ^{13}CH_3$ | $^{13}C$ & $^{13}C$ | 0.000119 | 0.000121 | -11.33 | -11.02 | 0.0166 |
| $^{13}CH_2D - CH_2 - CH_3$ | $^{13}C$ & D | 9.84E-06 | 1.01E-05 | -25.78 | -27.95 | 6.106 |
| $CH_3 - ^{13}CHD - CH_3$ | $^{13}C$ & D | 3.69E-06 | 3.36E-06 | 91.33 | 81.79 | 5.621 |
| $^{13}CH_3 - CHD - CH_3$ | $^{13}C$ & D | 7.21E-06 | 6.73E-06 | 69.27 | 76.63 | 0.4595 |
| $CH_2D - ^{13}CH_2 - CH_3$ | $^{13}C$ & D | 9.95E-06 | 1.01E-05 | -14.58 | -16.99 | 0.4072 |
| $^{13}CH_3 - CH_2 - CH_2D$ | $^{13}C$ & D | 9.78E-06 | 1.01E-05 | -31.87 | -33.87 | 0.0179 |

Table A.5: Equilibrium concentrations of up to doubly substituted propane at 0°C and at random distribution of isotopes (i.e., infinite temperature). The last three columns compare different measures of isotopic enrichment.

## A.6 Averaging of clumped heavy isotope effect due to different isotopologues

We compute the combined clumped heavy isotope effect by averaging the clumped heavy isotope effect of individual isotopologues according to how they would be distributed at infinite temperature (random distribution of isotopes among the isotopologues), which is an excellent approximation provided the $\Delta$-values are small.[95]

$$\Delta_{\text{D+D}}^{eq}(C_2H_6) = \frac{6\Delta_{CHD_2 - CH_3}^{eq} + 9\Delta_{CH_2D - CH_2D}^{eq}}{15} \tag{A.28}$$

$$\Delta_{^{13}C+D}^{eq}(C_2H_6) = \frac{6\Delta_{^{13}CH_2D - CH_3}^{eq} + 6\Delta_{CH_2D - ^{13}CH_3}^{eq}}{12} \tag{A.29}$$

$$\Delta_{\text{D+D}}^{eq}(C_3H_8) = \frac{6\Delta_{CHD_2 - CH_2 - CH_3}^{eq} + \Delta_{CH_3 - CD_2 - CH_3}^{eq}}{28}$$
$$+ \frac{9\Delta_{CH_2D - CH_2 - CH_2D}^{eq} + 12\Delta_{CH_2D - CHD - CH_3}^{eq}}{28} \tag{A.30}$$

$$\Delta_{^{13}C+D}^{eq}(C_3H_8) = \frac{6\Delta_{^{13}CH_2D - CH_2 - CH_3}^{eq} + 2\Delta_{CH_3 - ^{13}CHD - CH_3}^{eq}}{27}$$
$$+ \frac{6\Delta_{CH_2D - CH_2 - ^{13}CH_3}^{eq} + 9\Delta_{CH_2D - ^{13}CH_2 - CH_3}^{eq} + 4\Delta_{^{13}CH_3 - CHD - CH_3}^{eq}}{27} \tag{A.31}$$

$$\Delta_{^{13}C+^{13}C}^{eq}(C_3H_8) = \frac{\Delta_{^{13}CH_3 - CH_2 - ^{13}CH_3}^{eq} + 2\Delta_{^{13}CH_3 - ^{13}CH_2 - CH_3}^{eq}}{3}. \tag{A.32}$$

When the resolution of mass-spectroscopic measurements is not sufficient to distinguish between isotopologues with the same mass number, it is useful to consider the combined clumped heavy isotope effect $\Delta_{M+2}$. We report the calculations of these values by weighing the individual $\Delta_X^{eq}$ as if isotopes were distributed randomly.[95]

$$\Delta_{18}^{eq}(CH_4) = \frac{6r_D \times \Delta_{D+D}^{eq}(CH_4) + 4r_{^{13}C} \times \Delta_{^{13}C+D}^{eq}(CH_4)}{6r_D + 4r_{^{13}C}} \tag{A.33}$$

$$\Delta_{32}^{eq}(C_2H_6) = \frac{15r_D^2 \times \Delta_{D+D}^{eq}(C_2H_6) + 12r_{^{13}C}r_D \times \Delta_{^{13}C+D}^{eq}(C_2H_6) + r_{^{13}C}^2 \times \Delta_{^{13}C+^{13}C}^{eq}(C_2H_6)}{15r_D^2 + 12r_{^{13}C}r_D + r_{^{13}C}^2} \tag{A.34}$$

$$\Delta_{46}^{eq}(C_3H_8) = \frac{28r_D^2 \times \Delta_{D+D}^{eq}(C_3H_8) + 27r_{^{13}C}r_D \times \Delta_{^{13}C+D}^{eq}(C_3H_8) + 3r_{^{13}C}^2 \times \Delta_{^{13}C+^{13}C}^{eq}(C_3H_8)}{28r_D^2 + 27r_{^{13}C}r_D + 3r_{^{13}C}^2}. \tag{A.35}$$

Here the $r_{iso}$ stands for the molar fraction of the isotope (deuterium or carbon-13) in the sample.

# ADDITIONAL RESULTS

## B.1   Bulk fractionations

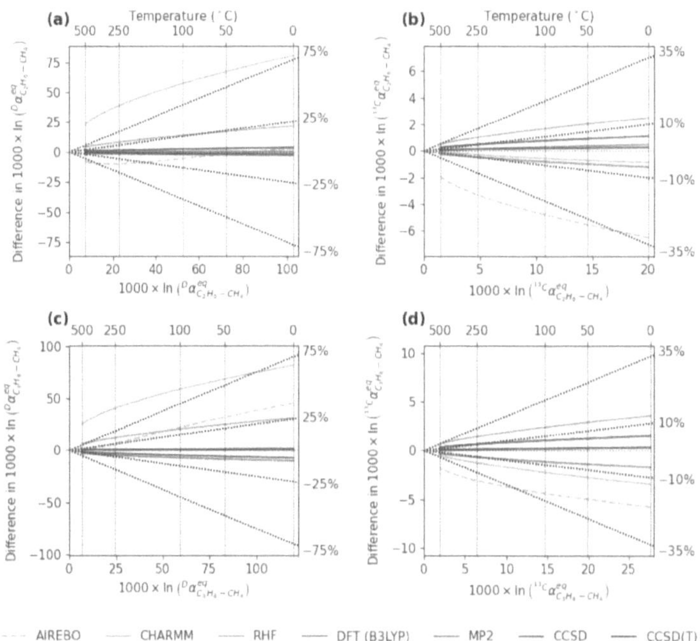

Figure B.1: **Comparing the differences in harmonic fractionation** relative to the F12/ATZ method presented in Fig. 6.2 (only aug-cc-pVTZ basis set results are shown here) relative to low-cost restricted Hartree-Fock (RHF) and empirical force fields (AIREBO, CHARMM) methods. The slanted black lines denote relative difference in fractionation and the dotted horizontal line is placed at y=0 where the reference (F12/ATZ) result would be.

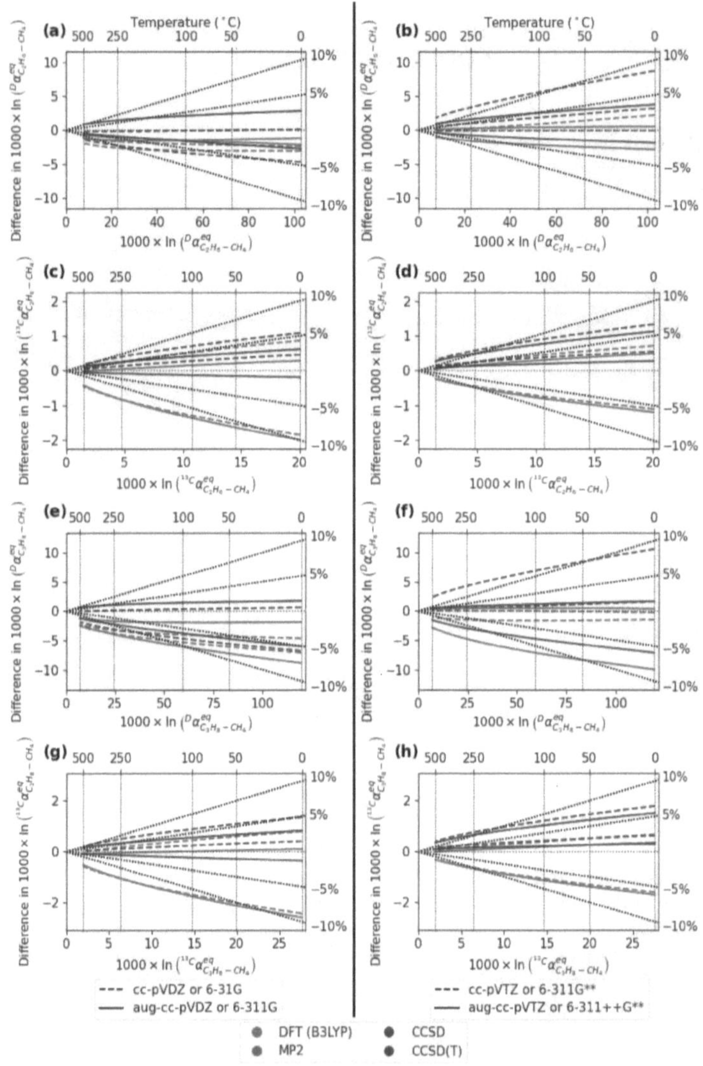

Figure B.2: **Comparison of the data presented in Fig. 6.2** (right panels) to analogous calculations with the smaller basis set size (left panels).

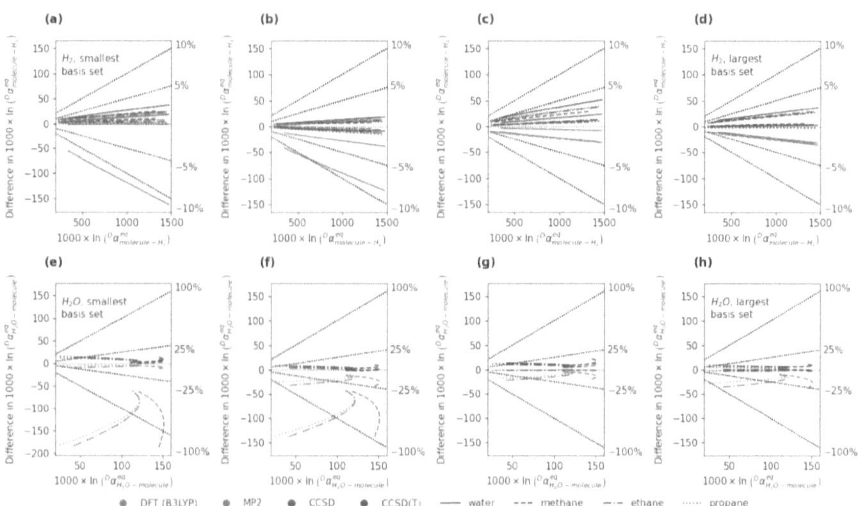



Figure B.3: **Comparing the results from Fig. 6.3** (repeated here on the four rightmost panels) to the same calculations performed with smaller basis set sizes: 6-31G for DFT and cc-pVDZ for other methods on panels (a,e) and 6-311G for DFT and aug-cc-pVTZ for other methods on panels (b,f).

| Poly order | $0^{th}$ | $1^{st}$ | $2^{nd}$ | $3^{rd}$ | $4^{th}$ | $5^{th}$ | Max Residual |
|---|---|---|---|---|---|---|---|
| Hydrogen | -119.624 | 353.6873 | 40.354 | -15.8056 | 3.034618 | -0.23047 | 0.054621 |
| Water | -200.912 | 501.4586 | 231.3391 | -74.4679 | 12.20933 | -0.80799 | 0.102199 |
| Methane | -80.8961 | 157.8892 | 420.5002 | -127.928 | 20.30959 | -1.32297 | 0.012594 |
| Ethane | -68.9123 | 126.3321 | 452.3712 | -137.124 | 22.24063 | -1.49596 | 0.07143 |
| Terminal | -78.7779 | 148.2027 | 423.7967 | -124.611 | 19.65664 | -1.29016 | 0.116424 |
| Center | -92.3679 | 182.6488 | 396.3042 | -105.781 | 15.06961 | -0.88493 | 0.183891 |
| Propane | -82.7114 | 158.1243 | 415.6881 | -119.348 | 18.40024 | -1.17987 | 0.122139 |

Table B.1: Fifth order least squares fit coefficients for **deuterium** fractionation and the site-specific isotope effect in propane. Last column contains the value of the maximum residual from fit.

| Poly order | $0^{th}$ | $1^{st}$ | $2^{nd}$ | $3^{rd}$ | $4^{th}$ | $5^{th}$ | Max Residual |
|---|---|---|---|---|---|---|---|
| Methane | -7.17911 | 16.79339 | 10.32542 | -2.05092 | 0.144118 | 0.002915 | 0.025833 |
| Ethane | -4.13426 | 8.570187 | 17.40372 | -3.74421 | 0.41383 | -0.01811 | 0.005173 |
| Terminal | -0.77945 | 0.73986 | 24.20712 | -6.58584 | 1.00921 | -0.06638 | 0.013485 |
| Center | 2.220193 | -7.05399 | 30.76189 | -8.04578 | 1.192091 | -0.07707 | 0.027543 |
| Propane | 0.219161 | -1.85393 | 26.38768 | -7.07111 | 1.070355 | -0.06999 | 0.010011 |

Table B.2: Fifth order least squares fit coefficients for **carbon-13** fractionation and the site-specific isotope effect in propane. Last column contains the value of the maximum residual from fit.

## B.2 Clumped isotope effects

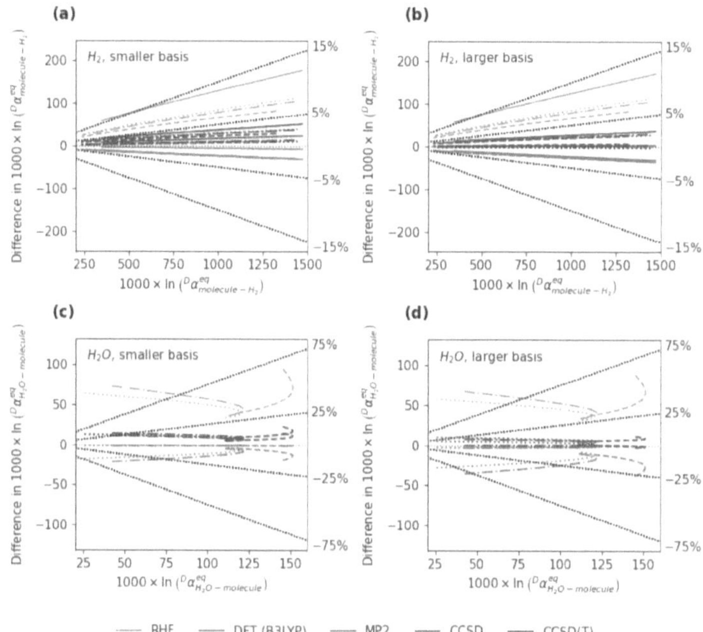

Figure B.4: **Comparing the differences in harmonic fractionation relative to the F12/ATZ method** presented in Fig. 6.3 relative to low-cost restricted Hartree-Fock (RHF) methods. The slanted black lines denote relative difference in fractionation and the dotted horizontal line is placed at y=0 where the reference (F12/ATZ) result would be.

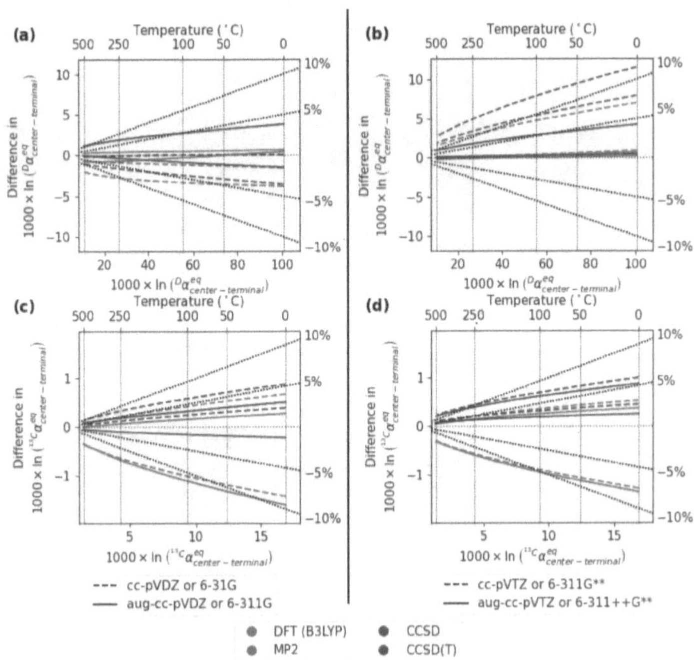

Figure B.5: **Comparing the results from Fig. 6.4** (repeated here on the right panels) to the same calculations performed with smaller basis set sizes (left panels).

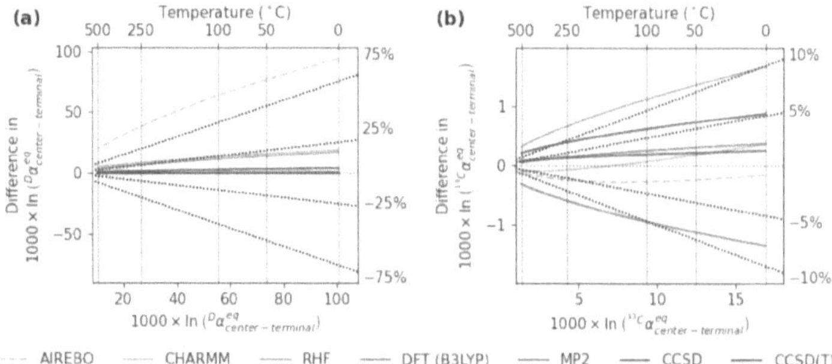

Figure B.6: **Comparing the differences in position-specific isotope effect of propane** relative to the F12/ATZ method presented in Fig. 6.4 relative to low-cost restricted Hartree-Fock (RHF), AIREBO and CHARMM methods. Only the largest basis sets we used (aug-cc-pVTZ and 6-311++G$^{**}$ for DFT) are plotted for the electronic structure methods. The slanted black lines denote relative difference in fractionation and the dotted horizontal line is placed at y=0 where the reference (F12/ATZ) result would be.

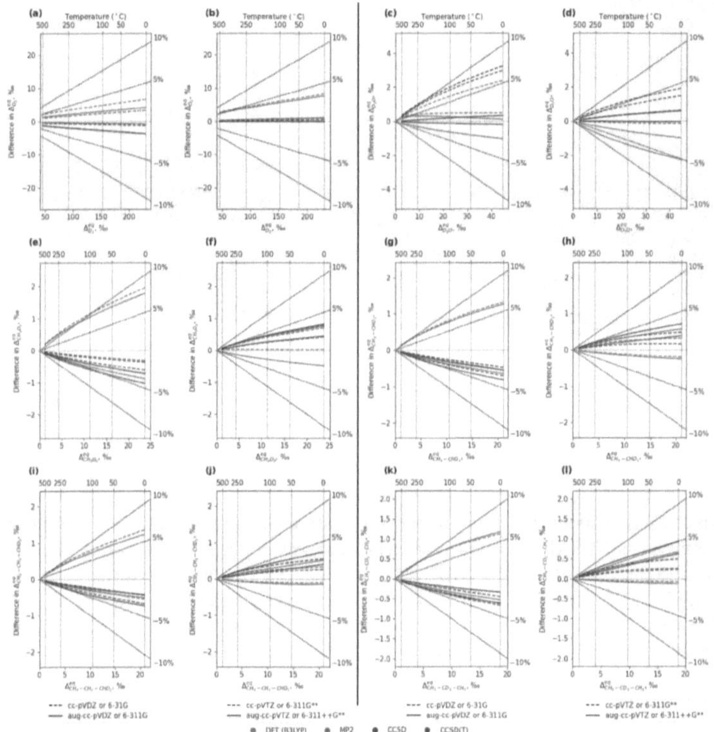

Figure B.7: **Comparing the results from Figure 6.5** (repeated here in the second and forth column) to the same calculations performed with smaller basis set sizes (first and third columns).

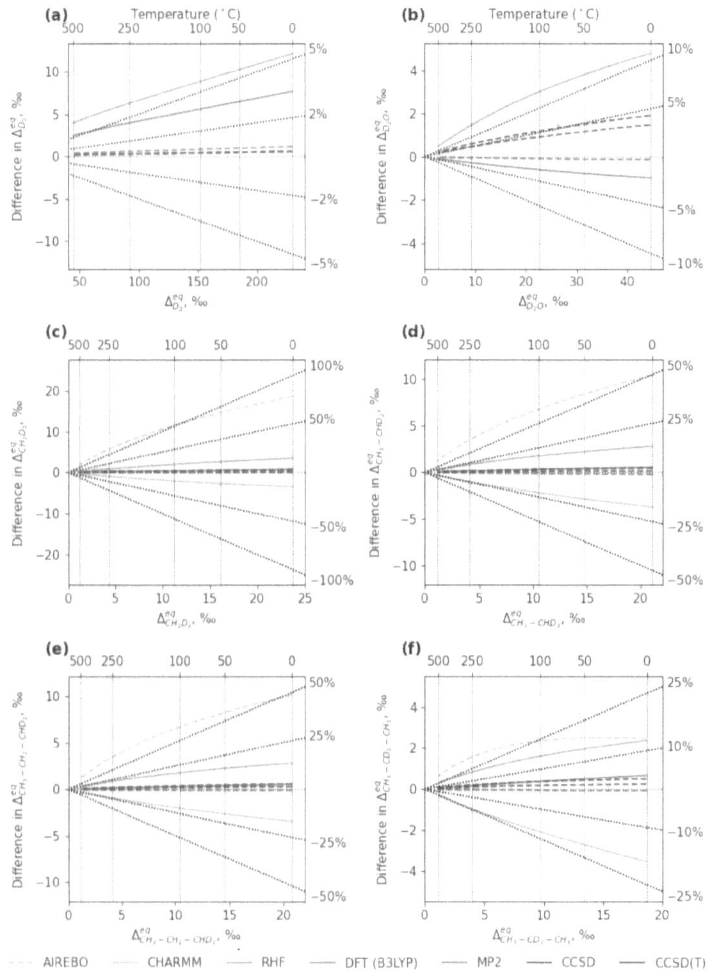



Figure B.8: **Comparing the differences in** $D + D$ **clumped isotope effect** relative to the F12/ATZ method presented in Fig. 6.5 relative to low-cost restricted Hartree-Fock (RHF), AIREBO and CHARMM methods (the last two for alkanes only). Only the largest basis sets we used (aug-cc-pVTZ and 6-311++G$^{**}$ for DFT) are plotted for the electronic structure methods.

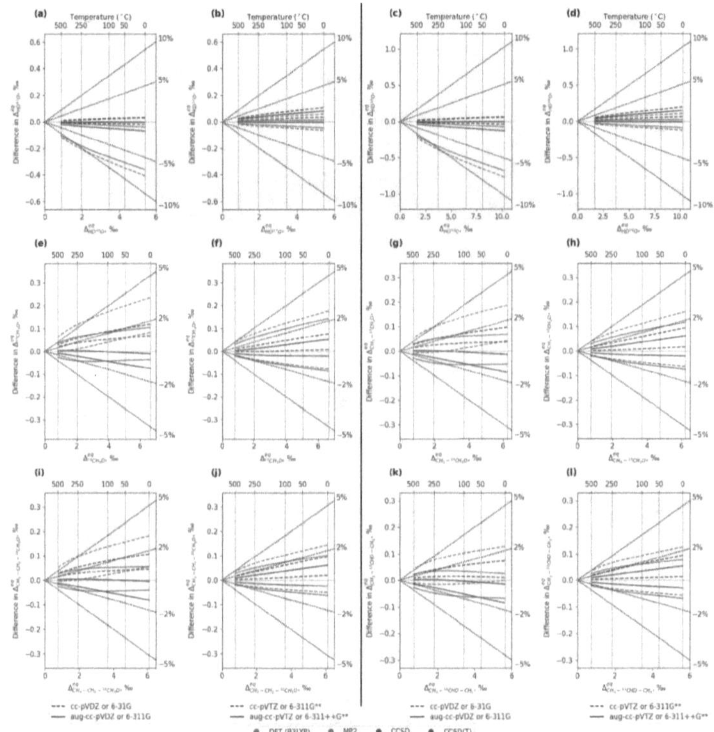

Figure B.9: **Comparing the results from Fig. 6.6** (repeated here in the second and forth column) to the same calculations performed with smaller basis set sizes (first and third columns). Note the larger relative deviation for clumped effect in water (panels a-d).

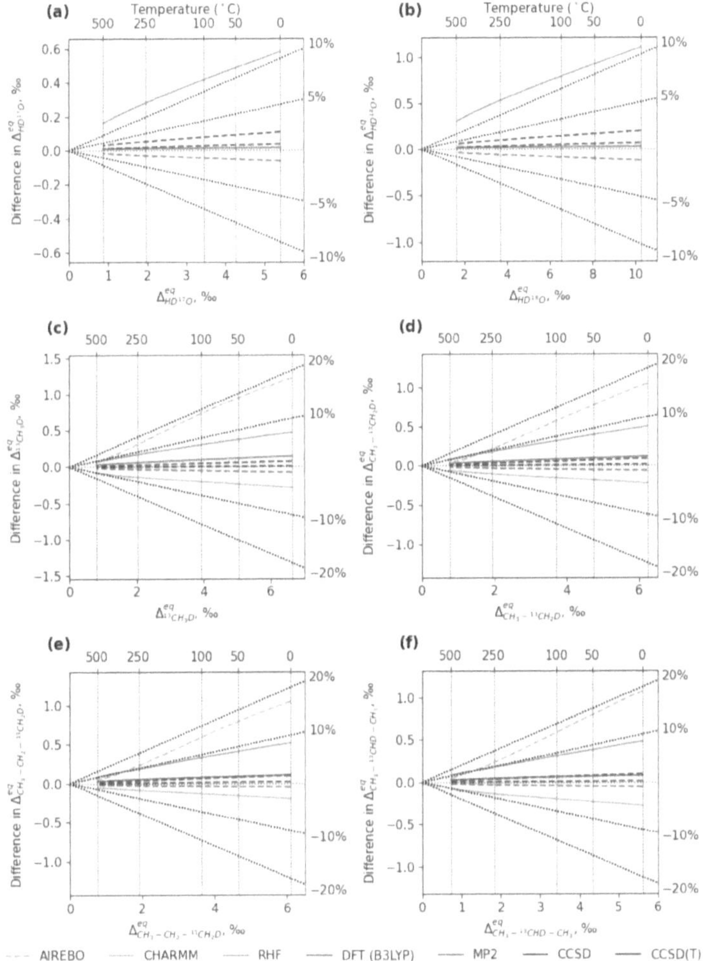



Figure B.10: **Comparing the differences in** $^{13}$C + D **clumped isotope effect** relative to the F12/ATZ method presented in Fig. 6.6 relative to low-cost restricted Hartree-Fock (RHF), AIREBO and CHARMM methods (the last two for alkanes only). Only the largest basis sets we used (aug-cc-pVTZ and 6-311++G** for DFT) are plotted for the electronic structure methods.

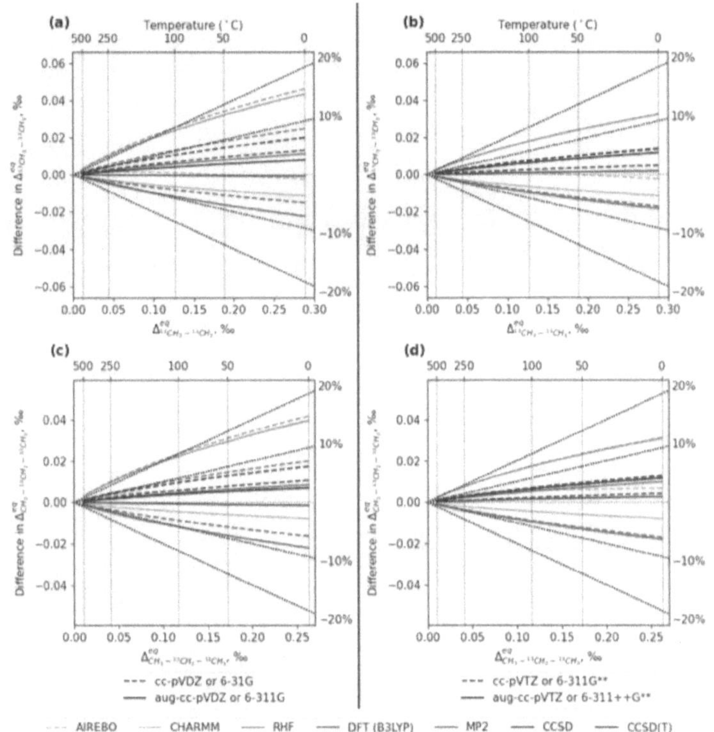

Figure B.11: **Comparing the differences in** $^{13}$C + $^{13}$C **clumped isotope effect** relative to the F12/ATZ method presented in Figure 3.6 relative to low-cost restricted Hartree-Fock (RHF), AIREBO and CHARMM methods (the last two for alkanes only). Only the largest basis sets we used (aug-cc-pVTZ and 6-311++G$^{**}$ for DFT) are plotted for the electronic structure methods.

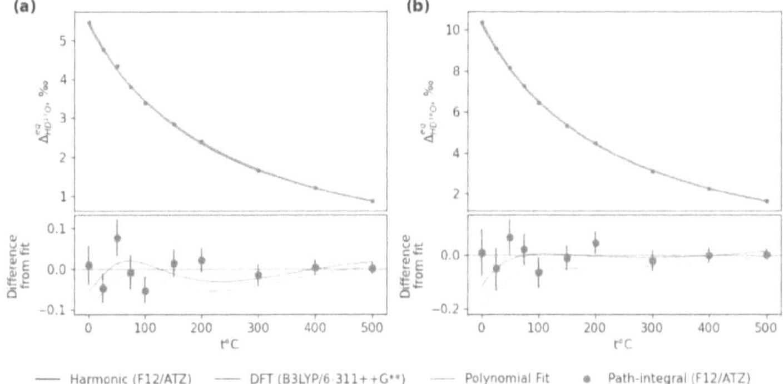

Figure B.12: $^{17}O + D$ (a) and $^{18}O + D$ (b) clumped isotope effects.

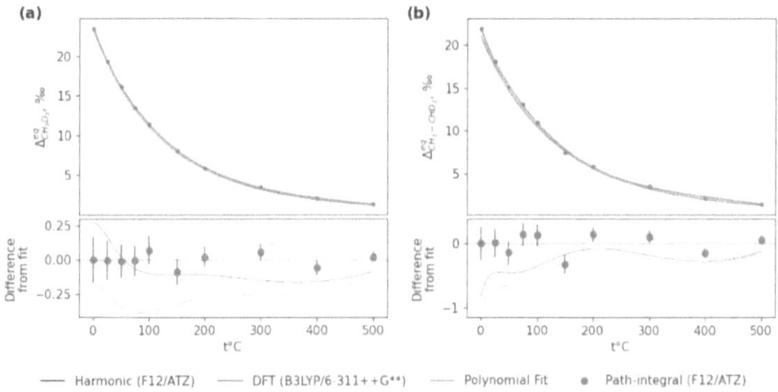

Figure B.13: $^{13}C + ^{13}C$ clumped isotope effects in methane (a) and ethane (b).

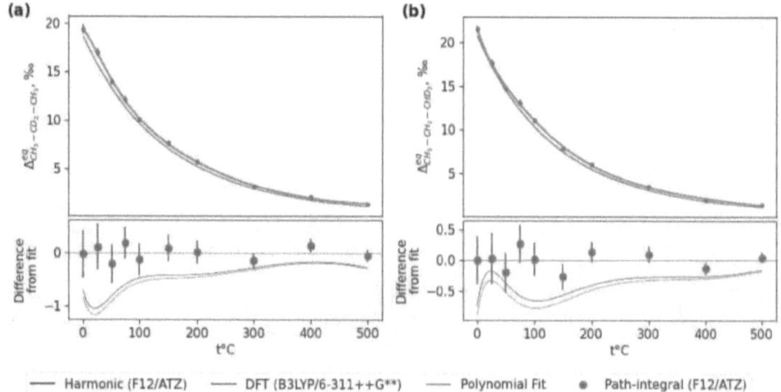

Figure B.14: D + D **clumped isotope effects in propane center (a) and terminal (b) positions.**

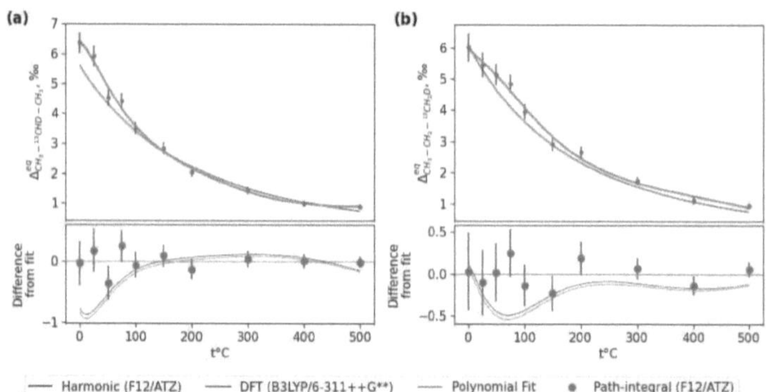

Figure B.15: $^{13}$C + D **clumped isotope effects in propane center (a) and terminal (b) positions.**